Chapter-2

Income and Inequality Among Rural Households in India and Punjab: A Review

The review for income inequality among rural household provided insight into the sources of income of farm households as well as the extent of inequality. Determining income of a farm household is a tedious task as lot of variation is observed in farm household income composition due to varying land holding size, cropping pattern, participation in subsidiary activities, family income, etc. The studies have been divided into two categories for ease of understanding i) income among and income inequality among rural households in India and ii) income and income inequality rural household in Punjab.

2.1 INCOME SOURCES AND INCOME INEQUALITY AMONG RURAL HOUSEHOLD IN INDIA

Thakur et al (2000) conducted study in rural areas of Bihar to analyse income distribution and poverty. To collect representative data a sample of 847 households was taken from eight villages from four different agro-climatic zones of Bihar. Gini coefficient was used to measure income inequality, a Gini decomposition analysis as developed by Fei, Ranis, and Kuo (1978) and Pyatt, Chen and Fei (1980) was applied to quantify the relative contribution of various income sources to the overall degree of inequality in household income. Multiple regression was used to identify factors that determine household income. Probit model was used to identify probability of individual being poor. Various means of livelihood recorded in the study were cultivation, livestock and fisheries, agricultural wage labour, services, trade and non-agricultural labour. Agriculture was source of income for 93.3 per cent individuals and non-agriculture was source of 70.3 per cent households. It was stated that income was unequally distributed with top 10 per cent had 25 per cent share in income. Rice cultivation contributed only 13 per cent of the household income, while about 55 per cent income was from non-agricultural activities. The income inequality from non-agricultural source was more and less in technologically advanced villages.

Singh (2013) conducted study on income and livelihood issues of farmers in Uttar Pradesh (UP). Primary data on cropping pattern, yield, income from farming and other sources was taken from a sample 3474 households distributed in nine agro-climatic zones of UP in 2012 using multistage stratified random sampling design. Income from agriculture was calculated by calculating gross value of agriculture (Either sold or self-consumed) and deducting the operational cost of cultivation (excluding the value of family labour). Non-agricultural sources like salary, wages, business, trade, etc were also explored. Average income came out to be Rs. 1.29 Lakh per annum per household. Share of agriculture, livestock/fisheries, industry and trade, wages, service and others was 51.19, 13.07, 1.95, 7.46, 16.46 and 9.86 per cent respectively. In different farm size categories, income per annum for marginal, small, medium and large farmers was Rs. 92913, Rs. 145145, Rs. 200027 and Rs. 361520. Share of non-agricultural income was the highest on marginal farms (48.03%) and decreased with increase in farm size. It was 21.90 per cent for large farms while share of agriculture and allied activities was 78.10 per cent. Wages were divided into sub categories: agricultural wages in the village, agricultural wages out of village, non-agricultural wages in the village, non-agricultural wages out of village, MGNREGS/ other public works and the highest share (71.41% of total wages) was of non-agricultural wages out of village. No other sources had more than 10 per cent share. It was concluded from the study that the agricultural income was meagre to survive so farmers especially marginal and small farmers opted for non-agricultural works. Various suggestion like increase in farm productivity, tapping potential of animal husbandry, evolving Public Private Partnership (PPP) model to promote rural industries have been suggested, training of rural youth for new and emerging services were demanded.

Birthal et al (2014) conducted study on income sources of farm households in India. The study was conducted using NSSO data 2003. The income categories have been classified into crop production, livestock, wages and salaries and non-farm business. In deriving results, it was found that 91 per cent households have reported income from agriculture with 41.4 per cent share of agriculture in in total income. Non-farm income has been reported by 14.2 per cent households and constituted second important source having 24.4 per cent share in total income. About 46 per cent farmers have drawn income from wages and salaries and contributed 20 per cent share in farm household income. Across different farm size categories, participation rate in crop production was almost same for all categories but share of income was 16.2 per cent only for semi marginal category (upto 0.5ha) which increased with increase in farm size and it was as high as 67 per cent on large category (>4 ha). Participation in livestock activity increased with increase in size of land holdings, wages & salaries and non-farm income activities decreased with increase in farm size similar trend was observed in share of these activities in total income. The regression analysis was carried-out taking dependent variable as 'share of a source of income in total income', and the explanatory variables included were landholding size, tenure status, land productivity, labour availability, access to farm and non-farm credit, age, sex, caste and education of the household-head. The results showed education has a strong effect on income diversification and farmers having access to farm credit (for crops as well as livestock production) were less likely to diversify towards non-farm activities. He suggested that income of small farmers can be enhanced by non-farm sector and intervention labour market. Development of animal husbandry has been recommended as it reduce income inequality. Policy to develop rural non-farm for sustainable employment opportunities has been suggested and farmers must be provided easy finance for shift to non-farm sector.

Chakravorty et al (2016) conducted study on income generation and inequality in India's agriculture sector with emphasis on land fragmentation. The data used for the study was obtained from National Sample Survey Organization (NSSO) study situation assessment survey of farmers conducted in 2003 and situation assessment survey of agricultural households 2013. The discussion was restricted to four income generation categories namely wages, net receipt from cultivation, raising of animals and non-farm income. They found that monthly per capita income varied from Rs. 3872 in Punjab to Rs. 736 in Bihar and income from cultivation also varied from Rs. 2311 in Punjab to Rs. 250 in West Bengal. It was concluded that land possession was the key variable in determining income from cultivation and it accounted for half of income inequality. It was also found that farming animals was had most rapid income growth in 2013 with 3.2 times increase in net income. Ordinary Least Square (OLS) regression was performed to determine significance of land in determination of income, income source and income distribution. Of all the variables taken only area of land owned and share of irrigated land were significant. Gini coefficient was used to estimate income inequality and coefficient for 2003 and 2013 was 0.63 and 0.53 respectively which depicted that inequality was reduced at national level.

Meena et al (2017) initiated an enquiry into the rural income inequality and its determinants in tribal dominated areas. The study was conducted in year 2011-12 in Jharkhand. The authors have emphasized the importance of using primary data to identify nature and character of income inequality within society and devised policy to improve the income distribution. The data was collected from two districts, one socio-economically developed and other backward. Tools like Lorenz curve, Gini coefficient, diversification index and regression analysis was performed on the collected data. Average income for Dubaliya, Hesapiri, Dumariya and Durgapur came out to be Rs. 14871, Rs. 9066, Rs. 9885, Rs. 6378. Sources of income in sample households consisted of crop, livestock, farm labour, non-farm work, salaried job, caste occupation, business, remittance and pension. Household obtained 48.54 per cent income from non-farm work. In regression analysis to determine factor affecting income, education, size of household, share of non-farm income and adoption of high yielding varieties had significant impact on households' income. The inequality was estimated to be the highest in labour class (Gini 0.55), followed by large farmers (Gini 0.50). It was stated that with change in individual component of income total distribution may change. Income from relatively equally distributed source would increase even income distribution. Per capita income and education level were sources of income inequality. Increase in education level will increase equality in the villages.

A survey conducted by NABARD, namely, "NABARD All India Rural Financial Inclusion Survey 2016-17" for the time period January to June 2017 in 29 states and 245 districts (Anonymous 2018a). There were 40327 rural household included in this survey. Various sources of income included wage labour (consisting of agricultural labour, skilled labour, non-agricultural labour, MNREGA activity), government/private service, cultivation (which includes crop production, plantation, orchard), other enterprises (which includes traders, shopkeeper, petty business, selling NTFP, micro enterprises offering service and manufacturing), livestock rearing (using animals for gaining economic returns) and other sources (which includes income from rent on building or land or interest from bank deposits or dividends and interest earned from investment). On an average, total income of rural households came out to be Rs 8059 per month. Income from wage labour was Rs 3504 per month with highest share in total income (43.00%). It was followed by government/private service (24.0%), cultivation (19%), other enterprises (8.0%), livestock rearing (4.0%) and other sources (2.0%). The households were distributed into decile class according to Monthly per Capita Consumption Expenditure (MPCE) of the household and income and consumption expenditure per month per household was given for different decile groups. The income increased over the decile groups with the lowest at Rs 1019 per month per household and the highest Rs 20548 per month per household. Similarly, consumption expenditure also varied between lower and upper decile groups. There was wide disparity among the income of agricultural households in different states of India. It was found that Punjab (Rs 23133) has the highest income per month per households followed by Haryana (Rs 18496), Kerala (Rs 16927). In the low segmented was for Uttar Pradesh (Rs 6669), Andhra Pradesh (Rs 6920) and Jharkhand (Rs 6991).

NSSO conducted Situation Assessment Survey of agricultural households in 70th round of NSS (Anonymous 2016a). The survey defines an agricultural household as unit that receives agricultural produce of value greater than Rs 3000 for past 365 days and having at least one member self-employed in agriculture as principal or in subsidiary status during 365 days. The income from cultivation was net value of main product and by-product after deducting expenses, income from livestock included net value of sale of product (milk, poultry, fish, wool, live animal, manure), income from wages and salaries include sum of wages and salaries of all the individuals in the households, income from non-farm business include net receipts of non-farm businesses in which household was engaged. It was found that the average monthly income of the agricultural households for year 2012-13 was Rs 6427 with the highest share of cultivation (47.9%), followed by wage/ salary (32.2%), livestock (11.9%) and non-farm business (8.0%). The disparity of the income among the agricultural households can analysed by arranging the income in ascending order and then distributing the households into equal groups according to MPCE. There was wide disparity among the lower and upper decile groups. The lowest decile group has income Rs 1293 per month per household and that for highest decile group was Rs. 22269 per month per household. Similarly, consumption expenditure also varied between lower and upper decile groups. It was Rs 2465 per month per household for lower decile group and Rs 15134 per month per household for upper decile group. Difference in the income per household and consumption expenditure per household was negative for lower two decile groups and it was low for third, fourth and fifth decile group showing income and consumption distress in the lower strata of the distribution.

Anonymous (2015a) the study was conducted by Agricultural Economics Research Unit (AERU), at Institute of Economic Growth (IEG) to analyse farmers' income in India. The study was based on the NSS 70th round Situation Assessment Survey (SAS). The Gini decomposition for income of agricultural household was calculated for India. The income components were used to calculate share of source in total income (Sk), source Gini (Gk) and Gini correlation (Rk) and elasticities of income change by source. Share of source in total income (Sk) was highest for farming (0.46), followed by wages/salary (0.31), livestock (0.14) and non-farm business (0.08). Overall, coefficient of income inequality was 0.56 which was considered as high inequality. Source Gini (Gk) was the highest for all the sources of income, which means that all the households were not getting their income from all the sources making the Gini-coefficient for the source high. Gini correlation (Rk) depicts the correlation between the distribution of the source and the overall income distribution. It was high for farming (0.80) and non-farm business (0.70) while it was low for wages/salary (0.66) and livestock (0.64). The elasticity was negative for livestock and wages/salary which indicate these were inequality decreasing factors while all other factors were inequality increasing factors.

A survey namely, "Indian Human Development Survey" (IHDS) was conducted by National Council for Applied Economic Research and University of Maryland, for the year 2004-05 and 2011-12 with the sample size of 27010 households. From IHDS data, panel data of 19831 households was used by Ranganathan et al (2016) to compare the change in income over the two study periods. The sources of income were categorized into agriculture, agriculture labour, casual labour, salary, business, remittance and others. Agriculture comprised of crop cultivation and livestock, agriculture labour derived income from agricultural wages, casual labour derived income from non-farm wages, salary included salaried employment, business included employment in business, remittance included income from member living away from home and other source comprised of income from pension, scholarships, welfare benefits, income from rent, etc. Comparative analysis was done to examine the change in income of the rural households between the study period. Source-wise the highest increase in income was observed in other sources by 24.8 per cent followed by business (18.4%), remittance (11.5%), agriculture (2.1%) and agriculture labour (1.5%). On the other hand, the number of households earning income from salary and casual labour decreased by 5.0 per cent and 1.3 per cent, respectively. This change shows that households are diversifying income sources from single source to multiple sources. On an average, annual income of rural households has increased from Rs 34247 to Rs 94749 between 2004-05 and 2011-12, respectively. The income distribution of rural households was analysed by dividing the households into five equal quintile groups. There was gradual increase in income from lower 20 per cent households toward the highest 20 per cent households. There was steep rise in income even between fourth quintile and fifth quintile. Overall, income inequality was high as Gini coefficient was 0.56. Decomposition of Gini coefficient was used to analyse the sources of income inequality. Share of source in total income inequality (Sk) was the highest for agriculture, followed by salary, casual labour, business and agriculture labour. Gini correlation was high for agriculture (0.78), salary (0.77) and business (0.67) while it was low for agriculture labour (0.15), casual labour (0.33) and remittance (0.49).

2.2 INCOME SOURCES AND INCOME INEQUALITY IN AMONG RURAL HOUSEHOLD IN PUNJAB

Kaur and Sekhon (2005) studied the income inequalities in rural Punjab with special reference to rice-wheat system over the period of 1970-71 o 1987-88. It was observed that in rural areas the income disparity among cultivators was higher than non-cultivators. It was also reported that income inequalities were reduced in medium farm families over the study period. It was stated that the inefficient farmers in this category become efficient over the time. It was also reported that disparities for small and large farm families increase in the same period. It was concluded that overall disparity was increased over the study period (Gini coefficient increased from 0.27 to 0.38). The authors studied the determinants of income disparities and found that farm size, productivity and use of modern technology were the main factor causing disparities in income among cultivator households in rural areas. The adoption of new technologies like High Yielding Varieties (HYVs) and fertilizer use was adopted evenly among different farm size categories till nineteens but in late on their access became costly causing lesser availability to small farmers as well as increasing costs and reducing net returns. On the contrary, the large farmers enjoy economies of scale which further led to widening of income disparities among small and large farmers. The negative farm size and productivity relationship would result into decreasing disparities but it was observed that the relationship in Punjab has become non-negative or positive.

Productivity of wheat was neutral in relation to farm-size of wheat in 1988-89 and became positive in 1998-99 in Punjab. It was opined that reverse tenancy as observed in Punjab may be due to unfavourable conditions for smaller farmers.

Vatta et al (2008) studied interhouse variation in rural employment and income among rural households in Punjab. A sample size of 315 rural households was selected. using multistage random sampling technique. Data was collected for the agricultural year 2005-06. The households were categorised into cultivating and non-cultivating. The share of crop income among cultivating households increased with farm-size. On the contrary, share of agricultural labour and rental income decreased with increase in farm-size. For all farming households, the principal source of income was agriculture. Marginal and small farm-size categories have more diversified principal sources of income while 80 per cent of the medium and large households had agriculture as principal source of income. It was reported that the income was highly skewed toward larger farm-size categories. The non-farm income was least skewed.

Vatta and Sidhu (2010) conducted study on rural non-farm employment, income distribution and poverty using micro level data. A sample of 315 households was selected using multistage random sampling technique. Data was collected for the agricultural year 2005-06. The income was distributed in different quintiles and it was found that the share of Rural Non-Farm (RNF) sector was important source of income, but the per cent share of RNF sector declined as we moved from bottom toward top quintile. The government job was the main source of income in bottom quintile while share of self-employed and private income was the highest in top quintile. the contribution of different sources to inequality was computed. It was found that farm income contributed most to income inequality (58.05%) followed by non-farm income (23.9%) and transfer income (14.5%). Incidence of poverty was the highest among agricultural labour households (75%) followed by rural non-farm casual households (43.9%). Incidence of poverty was 15.2 per cent in households with single source of income. It was 22.7 per cent among the households having two income sources while it was the highest among the households with three income sources (26.4%). It was concluded that RNF sector caused poor households to fall deeper into poverty while it helped richer to households to increase their income. Landless, marginal and small farmer households derived their major share of income from RNF sector. It was important source of income even in upper quintile groups. It was suggested to generate employment in RNF sector by providing education and skill to the poorer households to bring them above the poverty line.

Pavithra and Vatta (2013) studied role of non-farm sector in sustaining rural livelihood in Punjab. The study was based on primary data collected from 94 households in Punjab. The data was collected for the year 2011-12 using multistage random sampling procedure. The households were classified into four broad categories on the basis of non-farm employment i.e., agricultural, non-farm, transfer income and others. The share of agriculture in total income was 14.84, 36.2, 84.9,

84.2 and 84.1 for landless, marginal, small, medium and large farmers, respectively. The share of non-farm business was 64.0, 26.7, 7.0 and 8.5 per cent for landless, marginal, small and medium farmers, respectively. There was no larger farmer earning income from non-farm source in the study sample. The share of transfer income and other sources of income were randomly distributed among different farm sizes. The diversification index score was 0.89, 0.86, 0.63, 0.64 and 0.51 for landless, marginal, small, medium and large farmers, respectively. The overall value of Gini coefficient for rural Punjab was 0.51. The Gini coefficient for farm income as well as non-farm income was 0.68, showing highly unequal distribution within different sources of income. It was found that non-farm income sources reduce income inequality. Landless, marginal and small farmers were engaged in less remunerative activities due low education level. Large family size, higher dependency ratio, smaller land holdings and social backwardness were identified as important factors affecting non-farm income.

Singh et al (2013) conducted study on income, expenditure and poverty among the farmers in Punjab. The data for the study was used from the scheme namely, "A Study into the Economics of Farming and the Pattern of Income and Expenditure Distribution in the Punjab Agriculture" for the year 2010-11. A sample of 175 farmers was taken from three agro-climatic zones of Punjab. The gross farm income for the farmers in Punjab was Rs 5.81 lakhs whereas the net farm income was Rs 3.30 lakh. The average non-farm income for farmers was Rs 22765. The net family income was worked out to be Rs 3.53 lakhs. Average farm expenditure was worked to be Rs 2.28 lakhs. It was further analysed that the expenditure per hectare increased for different farm size categories. It was Rs 43257 and Rs 108614 for marginal farms and large farms, respectively. It was further found that small farmers spent more on feed while large farmers spent on land rent and hired labour. Thus, smaller farmers pursed livestock enterprise to enhance their income. Domestic expenditure was worked-out at household as well as per capita level. it was found that average domestic expenditure per household was Rs 1.24 lakh out of which share of food was Rs 73380 and non-food was Rs 50956. Per capita domestic expenditure was Rs 20792 out of which food items was Rs 12271 and on non-food items was Rs 8521. It was also found that large farmers spent more on food as well as non-food items as than rest of the farm-size categories. The incidence of poverty was also analysed. There were 34.78 and 20.00 per cent below poverty line farmers in marginal and small farm-size categories. It was concluded that large farmers benefited from the welfare schemes and these schemes must embrace smaller farmers to encourage inclusive development.

Kaur and Singh (2014) conducted study in rural Punjab to examine income of marginal and small farmers. Sample size of 650 households was taken from three regions classified on the basis of productivity. Data were collected for the agricultural year 2007-08. Income per annum per household was Rs. 63373, it was Rs. 48095 for marginal farmers and Rs.80129 for small farmers. Per capita income per annum was Rs. 11696, it was Rs. 8773 for marginal farmers and Rs.

14982 for small farmers. Sources of income were farm business income, milk and milk products, poultry, hiring out agricultural equipment, hiring out labour in agriculture, leased out land, salaries, pensions, remittances and other sources which included hiring out labour in non-agricultural sector and income from small business, like shop keeping. Distribution of household income was analysed and it was found that bottom 10 per cent households claimed only 3.68 per cent income while upper 10 per cent had larger share of 29.04 percent income. Distribution per capita income also had unequal distribution with lower 10 percent had 3.22 per cent share while upper 10 per cent had 31.62 per cent share. Gini coefficient was found to be 0.41 for marginal and small combined. The Gini coefficient was 0.23 for 0.21 for marginal and small farmers, respectively. It was concluded that there was positive relationship between farm-size and income level.

Vatta and Pavithra (2016) analysed income inequality and asset distribution among rural households in Punjab. Two different datasets were used to conduct temporal analysis on income inequality. The data for the study was collected for the agricultural year 2005-06 and 2011-12. It was found that inequality increased between the year 2005 and 2010. Gini coefficient was 0.46 and 0.49 for two study periods. There was high inequality in land ownership. The household members in poorest quintile group had higher illiteracy. Over the time, the share of the poorest quintile decreased and the share of the richest quintile in the overall income share. It was identified that farm income and transfer income were identified as inequality increasing sources. It was suggested that proper planning must be done to ensure egalitarian norms in the society.

Choudhary and Singh (2019) studied income inequality among rural Punjab. Using data from NSSO 70th round survey pertaining to the year 2012-13. In the survey, a sample of 725 agricultural households was selected from the state. The quintile distribution of households and their level of income brought out that there was huge difference between the income of bottom and top quintile groups. The average income of bottom 20 per cent households was Rs 3981 per month while the average income of top 20 per cent was Rs 74581 per month. The income from crops had the maximum share in total income among different sources except bottom quintile where the maximum share was wages and salaries. The value of Gini coefficient for the rural households in the state was 0.517. the source-wise distribution of income brought out that non-farm business income was the most skewed (Gini index=0.96), followed by wages and salaries (Gini index=0.88), livestock (Gini index=0.61) and crop farming (Gini index=0.60). In district-wise analysis, it was found that in Ferozepur district was the most unequally distributed (Gini index=0.63) while the income in Barnala was equally distributed (Gini index=0.25). Theil index was constructed to examine whether the inequality was within agro-climatic zones and districts or not. It was ascertained that for both districts as well as agro-climatic zones the inequality was among the groups and minimal inequality was observed between the groups. It was suggested that district-wise crop plans should be implemented to decrease the inter-district inequality.

In agro-climatic zone-wise decomposition of income, it was found that non-farm income has income inequality decreasing effect for central, western, undulating plain zone. Thus, contribution of non-farm sources in inclusive growth of rural Punjab cannot be overlooked.

Saini and Kumar (2020) analysed the income pattern and consumption behaviour among different agricultural households in Punjab 2015-16. A sample size of 150 farmers was selected from different agro-climatic zones of Punjab. Overall average annual income of the agricultural household was estimated at Rs 6.46 lakh in Punjab. Income from crops was the main source of household income (78.08%). It was followed by livestock (13.01%) and non-farm income (8.92%). Across different farm size categories, the share of non-farm income increased with increase in farm-size while the share of livestock and non-farm income decreased with increase in farm-size. Per capita annual income for different farm-size categories was worked-out at Rs 39597, Rs 63412, Rs 87040, Rs 163922 and Rs 240371 for marginal, small, semi-medium, medium and large categories, respectively. The income of large farm-size category was six times the income of marginal farm-size category farmers. Similarly, consumption expenditure was also studied for agricultural households in Punjab. Household expenditure included expenditure on expenses like food, education, electricity, clothing, healthcare, etc. and expenditure not included were household assets formation. Overall, average annual consumption expenditure was Rs 1.66 lakh for agricultural households. Per capita annual consumption expenditure for agricultural households was Rs 29247. Gini coefficient for the sample was 0.49. Lorenz curve depicted that top 10 per cent agricultural households had 36 per cent income share. The study highlighted that marginal and small farms derived large share of their income from livestock and non-farm income. These sources helped in decreasing income inequality among farmers in Punjab.

To conclude, it was found that the income for rural population in India and Punjab revealed a completely different picture. In India, the income of the rural as well as agricultural households vary a lot across different states. The main sources of income in rural area were income from crops followed by wages & salaries and livestock. Agricultural households in Punjab have the highest income across all the states in India. The marginal and small farmers were more dependent on wages and salaries and income from livestock whereas the larger farm size categories were more dependent on crop cultivation. The extent of inequality was also high in India and the Gini coefficient was worked out to be 0.53. Similarly, the extent of inequality in Punjab was also reported to be very high (Gini coefficient=0.52). Decomposition analysis revealed that the livestock and non-farm business were income inequality decreasing factors while income from crop cultivation was inequality increasing factor.

Chapter-3

Farm-size and Productivity Relationship in India and Punjab: A Review

The objective of farm-size and productivity relationship aims to answer the extent of inequality in productivity across different farm-size categories. If some pattern in productivity differential prevails, it would be expected to identify the factors causing these variations and give solutions to address them. The review on farm-size and productivity has been divided into two subheads i) Farm size and productivity debate in India ii) Farm size and productivity debate in Punjab.

3.1 FARM-SIZE AND PRODUCTIVITY DEBATE IN INDIA

Three observations were given by Sen (1962) in context of Indian agriculture. He broadly gave three observation. Firstly, when family labour employed in agriculture is given an imputed value in terms of the ruling wage rate much of the Indian agriculture seems unremunerative. Secondly, the profitability of agriculture increases with the size of holding, profitability being measured by the surplus (or deficit) of output over costs including the imputed value of labour and thirdly, the productivity per acre decreases with the size of holding. It is the third observation which states that there was IR between farm size and productivity in Indian agriculture. It was observed that the increased output was related to increased inputs. While the explanation for increased use of bullock labour was its indivisibility for a given farm, there was no explanation for other factors. The study highlighted that higher use of labour on family-based farms was the reason for the lesser profitability of larger farms and more productivity of smaller farms. It was mentioned that large farms are mostly Wage Labour (WL) based farms and small farms are Family Labour (FL) based farms. He explained to support his observation that in a labour surplus economy with no work opportunities for family members outside the farm the objective of FL based farm would be to maximize output by applying labour up to point of zero marginal product of labour (maximum total product). On the contrary, WL based farm would try to maximize profits by restricting the labour application up to point where marginal product of labour is equal to prevailing wage rate. Hence, the small farms have higher output per acre. But the profitability is lower because of higher inputs costs (mainly the family labour cost labour cost). He further argued that the problem must be viewed as a problem in system of farming i.e., WL based or FL based. He hypothesized that if cooperative farming is practiced (another system of farming), it will give higher output same as FL based farming and thus, capitalist farming should not be considered as progressive trend. At the end, he also criticized the imputation of market wage rate to family labour without further elaborating the criticism.

Rao (1963) discussed the farm size and economies of scale in India. He started the discussion by making the distinction between industrial and agricultural production process. It was stated that internal economies of scale in industrial economies arise due to two factors i) Specialization and division of labour. ii) Optimum utilization of indivisible equipment (immobile) and management. As machinery is immobile (confined to firm only) output has to be scaled up to its optimal utilization. Hence, these two factors tend to increase the size of plant in industrial production. On the other hand, agriculture involves biological processes which require time-bound operations at each stage of production. Therefore, specialization and division of labour has limited scope in agriculture. Further, it was pointed that land as a factor of production is completely divisible. Similarly, other factors of production like improved seeds, fertilizers, insecticides, etc. are also divisible. Thus, these factors can be used on land without any size hindrance. It was further stated that as far as optimum utilization of bullock and farm machinery is concerned, they can be employed on other farms because they are not attached to land (mobile). It was explained that despite mobility, these fixed assets may remain under-utilized due to institutional and psychological obstacles to adjustment (due to seasonal demand of these equipment). It was also mentioned that management and supervision in agriculture can be increased to limited extent and their optimum level may crop up at lower levels of farm size, especially on labour intensive farms. Another aspect of lack of entrepreneurial capabilities was highlighted as most of the farmers practiced subsistence farming. The study used data of economics of farm management for different states for the year 1954-55 to 1958-59 for different states in India. Cost, returns and net income were calculated using inputs like seeds, manure, fertilizers, irrigation charges, feed of bullocks and the depreciation on fixed equipment (bullock, implements, machinery and farm constructions). It was found that net income per acre was higher on smaller farms. It was emphasized that higher income is not as such due to size but due to intensity of cultivation which was possible due to higher irrigation, higher cropping intensity and larger application of family labour. He further highlighted that up to certain range of farm size, these efforts ensured minimum subsistence of family needs.

Mazumdar (1965) studied farm size productivity relationship using FMS data for the year 1954-55. He claimed that IR observed in peasant faming in underdeveloped is due higher use of labour on small farms but criticized the labour-based explanation given by Sen (1962). It was claimed that supply price of family labour differed from that of hired labour for different production periods. It was stated that the marginal supply curve of hired labour for small and large farms will be same in case both types of farm are using any amount of hired labour on their farm. On the other hand, marginal supply price for family labour cannot be same for small and large farms. Further, this higher use of labour on small farms was attributed to the difference in the marginal cost of family labour on small and large farms. It was explained with the help of phenomenon observed empirically which states that the supply curve of labour will be upward sloping, in the words of the author of this study "the function is of a kind that produces higher marginal supply price of labour (equal to daily wage-rate) as the number of work-days supplied increases". Further he stated that, supply price of family labour will be lower than hired labour as hired labour possess marginal product only whereas family labour accrues rent along with marginal product. Hence, the hired labour will have horizontal wage rate as the average labour supplied in the region. The supply curve of family labour will be lower than existing wage rate when supply of family labour is less than the average supply of labour in the region. When the volume of work of family labour exceeds the average labour work supplied, the supply price of family labour is above prevailing wage rate. Discussing the implications of the above supply curve, it was stressed that need of labour input in agriculture is not as regular as industry. Hired labour is used less often only to use family labour profitably or effectively. On the issue of higher supply price of labour for larger farms, it was presupposed that the potential supply of family labour per acre on large farms is less than their counterpart. Hence, supply price for large farms will be higher not only in busy period but even for slack period. Moreover, certain agricultural operations are concentrated and have to completed within few days. Hence the family labour will be fully employed and there will be need to use hired labour. This situation will be faced by large farms more often. This helps to explain the hypothesis of higher use of labour on small farms.

Sen's (1964) work was extension of explanation to IR relationship between farm size and productivity to his earlier paper. In this study, labour based, technique based and fertility-based explanation were given. The labour-based explanation stated that opportunity cost of labour is low for family-based farm and sociological factors do not let the wage rate go down beyond certain level. It leads to more application of labour on family-based farm than the wage-based farm. As most of the small farms are family based hence it leads to more productivity on such farms. He used hypothetical example to explain IR between farm-size and productivity.

Let probability of finding job is 1 and no employment is 0. The prevailing wage rate is w. If there is extreme case of full employment, the probability will be equal to 1(p=1) and if there is no employment opportunity, the probability will be zero (p=0). As long as p<1 and other inputs are constant, there will be lower output per acre for wage-based farm as labour will be applied with restrain (MPL=w); but on family-based farm, labour will be fully exploited (MPL=0). Secondly, it suggested that the opportunity cost of family labour in slack period will be low. Hence, the farmer will tend to use more of labour in slack season. Technique based explanation was based on diseconomies of large scale. It states that small business provides more effectiveness in management and supervision. Another explanation based on fertility of land was given which states that small farm's land have higher fertility than large farm's land. It was explained as faster subdivision of fertile soils due to more growth of population on them. It was due to population growth and migration. It was emphasized that multiple hypotheses maybe correct but it is important to identify which hypothesis is relatively important.

Khusro (1964) in his article returns to scale in Indian agriculture has beautifully explained the alternative measures of farm size and farm efficiency. He explained that farm size can defined in single input (say acreage) or in terms of output. He further elaborated merits and demerits of either method. It was pointed out that the acreage as input is relatively free from annual fluctuations and changes in composition; but acres require standardization as the land quality varies for different plots. Output is another measure of size measurement but it is difficult to compare these measures physically (weight or volume) as it ignores difference in value (e.g. different varieties of crops, low income crops v/s high income crops, etc.). On the contrary, the output can be measured in value terms. It enables comparison. The demerits include difficulty in inter-temporal comparison, error in imputed value of produce retained for home consumption. Hence, it is difficult to compare different farms if farm size is measured in terms of value of output. Similarly, farm efficiency can be defined in terms of output per unit of single input or output per unit of all inputs. Yield per acre as measure of farm efficiency suffers from same draw backs as output because the yield has to be measured in value terms. Moreover, it is measure against single input only i.e., land. Another measure of efficiency put forward was of cost per unit of output. It was defended that farmers would be interested in minimizing cost. The demerit with this measure is that both the inputs and outputs have to be given imputed value in many cases which depends upon their opportunity costs. It was anticipated that the farmers are interested in minimizing paid-out costs only and hence, the measure of paid cost per output can be an ideal measure of efficiency. But it also includes output and has same shortcoming associated with estimation of output as discussed above. The data used to explain that the relationship was obtained from farm management studies for seven states of India for three years i.e., 1954-55, 1955-56 and 1956-57. It was found that gross output per acre and farm business income declined with increase in farm size; whereas net profit per acre increased with increase in farm size. It was also observed that as the farm size increased the land revenue decreased which represented decrease in average quality of soil. If the acreage is corrected for soil fertility (acreage is standardized), gross output per acre and farm business income per acre remains constant with farm size. Net profits had positive relation with farm size with few negative signs. The alternative measure of efficiency i.e., paid out cost per unit of output (Cp/O) shows constant ratio (Cp/O) as output increased. Also, paid out cost per unit of income (Cp/Y) also remained constant with in size of income. It was concluded that uncorrected acreage cause phenomena of inverse relationship, but there exist constant returns to scale in Indian agriculture in studied areas.

Bhardwaj (1974) conducted a detailed study on farm size productivity using FMS data for West Bengal, Uttar Pradesh, Punjab, Madras, Bombay and Madhya Pradesh. It was stressed that production decisions using different resource (input) combinations depend upon the market prices considered as opportunity cost. Three categories of farmers were classified i.e., big landlords, self-reliant medium farmers and economically weak peasantry with different bargaining positions. The markets become interlocked through price and non-price links. Generally, the power is vested with dominant rural class with the use of multiple functions (landlord, moneylender, etc.) simultaneously in number of markets which increased exploitation of the weaker peasantry. Universal applicability of the assumption of profit maximization was also questioned. It was proposed that the objective of the producer depends upon resource position and the objective may be to optimize profit or gross output or farm business income or subsistence. From the analysis, it was found that there was no systematic relationship between yield per acre and land holdings for individual crops in physical terms. There was significant positive relationship for wheat-gram and cotton on irrigated farms in Punjab. Similarly, the hypothesis of land quality difference for different categories was rejected as there was no difference in crop-wise productivity. The highly productive farmers used better crop mix or cultivated land intensively. Selection of crop represents the quality of soil. Hence, the difference in productivity lies in cropping intensity and cropping pattern. It was found that cropping intensity was inversely related to farm size. Further, the motive of producer may vary for different farm-size categories. For very small farmers, there is motive of subsistence. The farmers in this category are unable to meet their consumption needs with the produce from its farm. They often take consumption loan also to fulfil the consumption needs of the family. Middle farmers are self-sufficient and produce for market. They may be categorized as profit motive category of farmers. It was pointed that the large holdings have low level of irrigation and they are dependent on hired labour hence, they mostly grow the crops which are sturdier and requires less labour input.

An article published by Rudra and Sen (1980) tried to end the ambiguity on the issue of farm-size and productivity and concluded their observations over the studies on farm-size and productivity debated in the decades of sixties and seventies. It was asserted that both the researchers never contradicted on views but were presenting the same argument. They suggested that policy makers should differentiate between capitalistic and cooperative farmers. They explained that in cooperative farms, the advantage of family labour can be fully realised while retaining the benefits of large-scale farming. Hence, government should promote cooperative farming.

Deolaikar (1981) studied inverse relationship between farm size and productivity using regional data from India for 272 districts. Regression analysis using OLS technique was carried-out to meet the objectives of the study. Fertilizer application was taken as proxy for technical change. The results confirmed that small farm sector enjoyed higher yield as compare to large farmers and yield advantage diminished with even reversed with technical change. Reduced importance of labour and increased importance of fertilizer was the reason for the change in the relationship. It was suggested that redistribution should not be promoted for area experiencing technical change and credit facilities for small farmers were recommended so that they can use new technology.

Bhalla and Roy (1988) worked on farm size and productivity relationship with the explanation of inverse relationship (IR) based on land quality. IR has implications in theory and policies. There are two explanations, one is under uncertain conditions the farmer apply more labour and other explanation considers difference in land fertility. Earlier work by Khusro (1964) demonstrated that adjustment in land quality diminish or eliminate the IR. It was hypothesized that if the IR remains after adjustment for land quality variable then it can be validated that small farms have higher productivity and its most like cause would be labour imperfection. The data was used from Fertilizer Demand Survey (FDS) and the sample size was 21500 farm households across India. The model used were log-log and semi-log models. The log-log model was:

$$\ln Y' = \ln Y/H = A + B \ln H + e$$

Where Y=output and H=farm size (net cultivated area)

It has been assumed that farmer has adjusted to local environment by making relevant choices and understanding exogeneous non-choice determinants of farm productivity (e.g., weather, soil, etc.). Further, the technology and factor prices faced by the farmers are identical. Hence, an estimate of b is unbiased and should be equal to zero. Difference in output can result from three factors cropping intensity, yield difference of various crop, cropping pattern. Composition of output and difference in cropping intensity are ignored. Semi-log form was used as:

$$Y' = Y/H = A + B \ln H + e$$

When the farms were categorized on the basis of family labour, intermediate and hired labour based, the marginal product was the highest on farms with hired labour and the marginal product curve flattened for family farm. Thus, suggesting semi-log relationship between Y/H and H. In semi-log form the elasticity of output per acre with respect to farm-size was not constant but dependent on output per acre. Another model was formed using the land fragmentation as another variable. The relationship was estimated at four levels; district, sub-zonal, zonal and state.

IR was found for 17 states. Similarly, for zones level classification, 54 out of 78 agronomic zones have negative correlation. For sub-zonal classification, out of 142 sub-zones 73 sub-zones reports. At district level classification, out of 176 districts, less than 50 per cent districts have IR. When soil quality and exogeneous irrigation are incorporated in agro-climatic zones, the no of cases of IR drops in all the classification levels. It was found that smaller the unit of estimation i.e., the more valid the assumption of homogeneous environment, lower the number of samples reporting a negative relationship for traditional specification. Moreover, the addition of soil quality variable always decreased the number of cases showing IR. The study indicates that there was no homogeneity in the states and these heterogeneous differences were the major factors explaining land productivity differentials.

Frisvold (1994) purposed the hypothesis of complete labour homogeneity. He used plot level rice production data from village level studies conducted by International Crops Research Institute for the Semi-Arid Tropics (ICRISAT) in Mahbubnagar in Andhra Pradesh for the four years from 1981 to 1984. Cobb-Douglas model was used for testing the proposed hypotheses. It was found that the productivity of hired labour increases with supervision. He suggested the use of sensitivity analysis in future studies to check for robustness of supervision response function when elasticities were low.

Sengupta and Chattopadhyay (1997) studied farm size and productivity relationship using disaggregated data. He has extensively cited paper by Bhattacharya and Saini (1972) to solve farm size and productivity riddle. Farm level disaggregated data was obtained from "Comprehensive Scheme for Studying Cost of Cultivation" for West Bengal pertaining to year 1989-90. Entire area was divided into six agro-climatic zones and sample of 600 farmers was taken by probability proportion to cultivated area. Regression analysis, Kendall's rank correlation coefficient and Chi-square test. Standard analysis of variance was used to test suitability of regression model. It was concluded from the study that inverse farm size relationship existed and even became stronger in agriculturally developed regions of West Bengal. The author advised further studies at different parts of country to consolidate the claim.

Lamb (2002) examined the role of land quality and imperfect markets in inverse farm size productivity relationship. He used the data of International Crops Research Institute for the Semi-Arid Tropics (ICRISAT) for 10 years and the sample size was 1060 households. It was found that land quality and market failure explain most of the inverse relationship in random affect estimates while measurement error in the farm size do explain inverse relationship in fixed effects especially. It was suggested that fixed affect models should be cautiously used in in developing countries where tendency to change farm size over time is minimal.

Singh et al (2011) worked out relationship between farm size and productivity with special emphasis on strengths of small holders. The study was classified into four sections including first, global landholding scenario, second, distribution of land holdings in India, third, farm-size category wise change in input use, improved technology and crop intensity in last three decades along with the farm size and productivity relation and at last, strengths of smallholders. It was found that input use like fertilizer, High Yielding Varieties (HYVs) and access to irrigation has been more on small farms as compared to large farms and the pattern has been intact for past three decades. Cropping intensity was also higher on small farms. Farm size and productivity relationship was analyzed using NSSO Situation Assessment Survey in year 2002-03. Value of crop per hectare was the highest (Rs. 25173) on 0.4 ha category and least (Rs. 7722) on large farm. Thus, even with increase in use of advanced technology favouring scale production, marginal and small farmers have performed better, inverse relationship between farm size and productivity was reported. But despite higher productivity per capita income was lower on small farms due to lower land-man ratio. It provides support for decreasing land inequality and lowering land ceiling limit to increase productivity.

Deininger et al (2016) conducted study using 'Additional Rural Incomes Survey' & 'Rural Economic & Demographic Survey' (ARIS/REDS) panel survey data collected by National Council for Applied Economic Research (NCAER) for 242 villages in 16 states of India for years 1982, 1999 and 2007. The panel data was used to study farm size and productivity relationship and change in farm size and productivity relationship overtime. In Descriptive analysis, weakening of IR between farm size and productivity over time was observed. Regression Analysis was used in which dependent variable was value of total crop output while independent variables were total cropped area, vector of household characters, land quality and year dummy. The coefficients of farm size and productivity were negative for both periods 0.73 and 0.95 for data for period 1999 and 2007 suggesting that magnitude of relationship has weakened over time. The labour market was considered imperfect if the amount of labour used in agricultural production by farm household depends on the labour endowment. It was concluded that labour market imperfection, existence of efficiency difference between hired labour and family labour also leads to IR in India.

Gollin (2018) analysed the existing literature on farm size and productivity at consultation, focusing theme "Agricultural and Rural Development Research and Investment on Achieving Sustainable Development Goals". He emphasized the need to study farm size and productivity relation citing small size of farms in developing country and suggested need of land reforms to change the situation or these are efficient. He mentioned ambiguity in defining productivity and farm size. He analysed output: worker ratio which was equal to output: land (yield) and land: labour ratio and found that there are huge differences in different countries for output per unit of labour as compare to output per unit of land. It was also mentioned that inverse relationship was found in GPS estimates also. But he stressed that even if inverse relation existed, farm size is not important variable in explaining higher crop yield and output per unit labour may be higher on larger farms. In the concluding remarks he pointed that even in the presence of IR small farms were unable to provide decent income even if yields were increased to world level and suggested developing new markets for consumption of agriculture goods, non-farm employment and urban growth.

Singh et al (2018) used plot wise panel data from project "Tracking Changes in Rural Poverty in Households and Village Economies in South Asia" to study income of smallholders of Bihar. A sample of 160 respondent farmers were selected from four selected villages. An equal number of farmers were selected representing different farm size categories. Paddy and wheat crops were the principal crops of the region. Fertilizer consumption was the highest on larger farms and subsequently productivity was also the highest on larger farms. Regression analysis brought out that there was positive relationship between farm size and main output. Higher productivity on large farms was due to intensive use of modern technology and opposite was true for small holders due to prevalence of uneconomic land holdings which lead to part time farming and capital constrains. Policy implication from the paper includes development institutional financing system, agricultural extension network and farm technology centre.

3.2 FARM-SIZE AND PRODUCTIVITY DEBATE IN PUNJAB

Rao (1967) in the study size of holding and productivity tried to verify the inverse relationship between farm size and productivity using farm level disaggregated data. The data was collected by Agro Economic Research Centre (AERC), New Delhi in selected villages of Punjab and Uttar Pradesh. It was found that productivity remained constant over whole range of holdings. Irrigated area as well as other expenses on seeds, fertilizers and manures also remained constant over different farm size. Same results were obtained with gross and net area. It was concluded that when there is no difference in proportion of area irrigated among different size class output tends to remain constant over all farm size.

Rudra (1968a) in his study farm size and yield per acre extended the work of Rao (1967) with 20 villages and 1198 farms in parts of Punjab, western U.P and Haryana. Two variables i.e., size of farm in acres and yield per acre in value was compared. Size of farm was sum of cultivated and uncultivated area. Productivity was calculated with gross output as numerator and gross cultivated area as denominator. It was done to treat extent of cropping intensity as distinct factor affecting the economics of farming. The value of output was also calculated somewhat differently. Modal price of village was taken to remove pseudo price fluctuations causing arbitrary increase or decrease in value of productivity. different methodology of checking association was adopted instead of regression approach taken by earlier researchers. It was done by using F-test for one-way classification by grouping data in 0-5 acres, 5-10 acres, 10-15 acres, 15-20 acres, 20-50 acres and more than 50 acres. The study found that in 2 out of 20 villages farm size and yield were significant i.e., the variables showed association. The villages which were significant were analysed further. There was systematic increase in yield per acre up to 20 acres which decreased afterwards in one village. The other village had violent ups and downs showing no systematic pattern. Hence, the author rejected the inverse relationship hypothesis for this region.

Rudra (1968b) worked on return to scale in Indian agriculture. In the earlier paper Rudra (1968a), he rejected the prevalence of inverse relation in Indian agriculture which intensified the debated on the topic. In the present study, he used FMS data to verify its results. The data was average data of each group (aggregated data). Therefore, rank correlation coefficient was used to analyse the correlation between farm size and productivity. It was discussed that the negative correlation implied decreasing returns to scale (inverse relationship), positive correlation implied increasing returns to scale and no correlation implied absence of any of the two patterns. The results for individual crops showed significant negative correlation in Madhya Pradesh and Uttar Pradesh. For Punjab, the correlation was positive. The farm level data for 17 districts was analysed and it was observed that only 9 out of 17 districts had negative correlation. Hence, it was suggested to restrain from generalizing the phenomena of IR between farm size and productivities in India.

Bhattacharya and Saini (1972) analysed disaggregated data to know the farm size and productivity relationship. They choose two districts of green revolution area namely, Muzaffarnagar and Ferozepur in Uttar Pradesh and Punjab respectively. The data was taken for fifties and sixties to examine effect of green revolution on farm size and productivity. The relation between farm size and value of output per acre on the one hand and as gross cropped area and value of output per acre of gross cropped area on the other hand were analysed. The Pearson correlation was conducted for each village separately and followed by Kendall's rank correlation. Further, effect of aggregation over villages was investigated by using ANCOVA tests of homogeneity so as to ensure that size-productivity relationship for different villages could be specified by using regression equation. In Ferozepur, correlation was weak and non-significant except for 1956-57. Thus, farm size and productivity relationship was not clear. In Muzaffarpur, the coefficient was positive and significant for all years. Overall aggregation was inefficient as it made coefficient larger than actual. It was also mentioned that value of intercept explained more variation in productivity than farm size or gross cropped area. They asserted that even if conducting separate village-wise analysis is not possible, it would be compensated by introducing dummy variable representing change of intercept for different villages. If dummy variable is not introduced the relationship obtained will be distorted. Other concluding remarks include use of other variables like farm business income per acre and carrying separate analysis for different crops.

Chattopadhyay and Rudra (1976) conducted study titled "Size-productivity revisited". In the study he beautifully explained aggregation effect. He gave example that if two villages have different productivity and the aggregation may lead to spurious results. On the other hand, if two villages have same land productivity but different average farm size, it would also lead to spurious results. The data for the study was collected from Tamil Nadu, Andhra Pradesh, Kerala, Uttar Pradesh,

Punjab, Madhya Pradesh, Orrisa, West Bengal, Assam and Gujrat. Rank correlation coefficient was used to work out relationship between output per acre and farm size. It was found that there was IR between output per ha and farm-size in all states under study except Punjab, West Bengal and Gujrat. There was IR between cropping intensity and farm-size. Also, IR was observed between irrigation availability and farm-size except in Kerala. Total input use was negatively associated with farm-size in all states except Punjab. Different reasons for high input usage behavior of small farmers were given like easy availability of family labour, indivisibility of capital, superiority of land (where applicable), better management and subsistence needs. They concluded that high input use in case of small farmers showed distress rather than efficiency. Further they finalized that the results could not be generalised the overall India agriculture.

Chadha (1978) used Farm Management Survey (1956-57 and 1969-70) and Agricultural Census (1970-71) to analyze farm size and productivity relationship in Punjab. The state was divided into three homogeneous regions formed on the basis of soil conditions, cropping pattern and irrigation system. He formulated various hypotheses for factors affecting farm size and productivity relationship, which included land quality, tenancy, irrigation, cropping intensity and cropping pattern and formulated a new hypothesis in which he asserted that net output effect was not determined by absolute expansion of either labour or capital but by relative expansion of both. Capital to labour ratio was analysed for 1969-70 data for different farm size categories for three different agro-climatic regions of Punjab. The stock of implements and machinery was divided into traditional and improved category. It was found that small farmers had traditional implements while medium and large farmer had edge with improved implements-machinery. Region II was the most advanced and region I was the least in case of per acre investment in implements-machinery and irrigation. Working charges were added for machinery use and irrigation for all users whether owners or non-owner and per acre expenditure was worked by adding fertilizers, manures, seeds, pesticides, etc. to arrive at "capital services" input. Capital to labour ratio was obtained by dividing "capital services" by human labour per acre. In different farm size categories, region II represents the highest capital-labour ratio followed by region III and the least in region I. In analysing farm size and output per acre relationship using log-linear regression, he found that inverse relationship disappeared and small farmers had clearly lost their edge in region II where increase in capital: labour ratio has taken place on large farms. It was suggested that small farmers must be provided with institutional machinery through government pool, cooperatives or private machine operations.

Ghose (1979) worked on the farm size and productivity relation using farm level data. The data was obtained from Farm Management Surveys from Hoogly district of West Bengal for year 1955-56, 1956-57 and 1971-72 and similarly, from Ferozepur district of Punjab for year 1955-56, 1956-57 and 1968-69. It was pointed that small farms used more labour per acre and more irrigation and manure was available on small farms. Thus, primitive technology and insufficient development of markets was prime factor for better performance of small farms.

The relationship between farm size and productivity changed from negative to positive in technologically progressed district (Ferozepur) while relation between farm size and labour intensity remained inverse. It was asserted that with technological progress like fertilizers, irrigation techniques, superiority of small-scale production would vanish.

A study on income distribution in Punjab was conducted by Bhalla and Chadha (1982) and they also analysed the farm size and productivity relationship. In the study, it was mentioned that in Pre-green revolution period there were significant difference of cropping pattern among different farm size groups and after 1965, cropping pattern has changed drastically, especially in Punjab in favour of wheat-rice crops. IR between farm size and productivity was observed due to more use of draught animals (48.63%) by marginal and small farmers. they analysed farm size and productivity relationship for wheat and rice crop separately. In case of wheat, they found that yield of wheat crop was size neutral as there was no increase or decrease in yields with increase in farm size. Yield of rice crop was less for marginal and small farm size category. It might be due to technological disadvantage of marginal and small farmers. It was observed that high yield of wheat and rice and the state policy to stabilize their prices have made wheat and rice lucrative crops in the state. Adoption of these crops was size neutral i.e.; the crops were adopted equally by all farm size category farmers.

Singh et al (2000) analysed the farm size and productivity relationship in their study namely, "Changes in Structural Distribution of Land Holdings and its Impact on Productivity of Major Crops in Punjab". The data was obtained from three agro-climatic zones of Punjab for five crops i.e., wheat, paddy, cotton, sugarcane and maize from the scheme of "Crop-Cutting Experiments" of Department of Agriculture, Punjab. Log linear regression was used to study the relationship between farm size and productivity for individual crops. It was found that productivity of paddy was size neutral. Farm size and productivity relationship was significant and positive for maize and it was non-significant in case of cotton crop. For wheat crop, relationship was positive in zone I only while it was non-significant for zone II and zone III. In case of sugarcane, farm size and productivity relationship was non-significant. It can be concluded that there was either neutral or positive relation which was due to diffusion of technology among all the farm size categories and hence pointing the need for large sized operational farms. Policy implications were given for abolishing of ceiling on the operational holdings so as to ensure capital formation, resource use efficiency and more productive farming.

To conclude, it was found that country as a whole and Punjab region explained opposite results for farm size and productivity. In India, most of the researchers reported prevalence of inverse relationship between farm-size and productivity. On the contrary, it has been reported that in last decade of twentieth century the there was no inverse/negative relation between farm-size and productivity in Punjab agriculture. Crop-wise analysis depicted either mom-significant relationship or positive relationship between farm-size and productivity in Punjab.

Chapter-4

Concepts and Methodology

A scientific research uses standardized methodology and statistical procedures which enhances the reliability and facilitates smoother application of the findings. The chapter on material and methods defines the locale of the study, data sources, basic concepts and analytical procedure used in the present study.

4.1 LOCALE OF THE STUDY

The present study was conducted in Punjab, a land-lock state bounded on the west by Pakistan, on the north by Jammu and Kashmir, on the north-east by Himachal Pradesh and on the south by Haryana and Rajasthan. The word Punjab is a combination of two Persian words, 'panj' and 'ab' signifying the land of five waters. Punjab holds a place of pride among the Indian States for its outstanding achievements in agricultural development. It comprises only 2.94 per cent of the cultivated area of country and contributed 5.28 per cent towards the total food grain production (Anonymous, 2019). State has earned a name of 'Granary of India' as it was the major contributor to central pool for buffer stock of rice and wheat with 45.3 per cent and 73 per cent contribution in 1980-81, respectively. Even in recent years, Punjab has contributed significantly to buffer stock with 25.53 per cent and 35.46 per cent of rice and wheat to central pool for the year 2019-20 (Anonymous, 2019). It was consequent to the high use of chemical fertilizer, pesticide, high yielding varieties, machinery and supported by extension of irrigation network, rural link roads, rural electrification, establishment of focal points and agricultural market centers, efficient delivery system of credit and other agricultural inputs along with effective implementation of agricultural price policy for wheat and paddy.

4.2 DATA SOURCE

The data were obtained from National Sample Survey Organization (NSSO) 70th round 'Situation Assessment Survey of Agricultural Households'. The data pertain to the agricultural year 2012-13. The number of sample villages and number of households surveyed by NSSO in Punjab was 94 and 725, respectively. There were 558 farm households among the sampled households.

4.3 IMPORTANT CONCEPTS AND TERMS USED IN THE STUDY

The study was conducted using household as basic unit of analysis. The concepts and terms used in the present study are briefly discussed below:

4.3.1 Farm household

A household is considered to be a farm household when at least one of the members is actively engaged in crop production. According to NSSO, a household can be defined as a group of people who share food regularly, and usually live in the same house or group of houses.

3.3.2 Farm household income

A farm household earns its incomes from various sources. The total income of the household was obtained by summing-up the following sources of income:

A) **Income from cultivation:** It is the income earned from cultivation of crops during the survey year. It also includes the value of the by-products. The total cost was subtracted from the total value of crop output to arrive at the income from farming. The inputs cost included were the paid-out cost of seeds, fertilizers, manures, plant protection chemicals, human labour, electricity, irrigation, diesel, repairs & maintenance, cost of hiring machinery, rent for leased-in land, interest on loans for cultivation purposes and other expenses.

B) **Income from livestock:** This is the income a household earns from sale of various livestock products. It is calculated after deducting the cost incurred on livestock production from total value of the produce. The costs incurred will include cost on maintenance of animals such as animal feeds, veterinary charges, interest, labour charges and other expenses.

C) **Income from regular wages and salary:** The income derived by any household member employed anywhere other than own household enterprises can be defined under wages and salaries. The individual receiving payments on regular basis are considered in this section.

D) **Income from casual wages:** The income earned by any household member by casual employment outside the household enterprises was defined as income from casual wages.

E) **Income from non-farm business:** This is the income derived by any household member by engaging himself in non-farm businesses activities such as hotels, restaurants, transport, construction and manufacturing. The collected information was related to expenses, output and net receipt of up to five major non-farm businesses of the households.

F) **Total income of farm household:** Total income of the household is obtained by summing-up the net income from above mentioned sources. Income from crops and livestock are estimated by deducting paid-out costs from gross returns. Net farm business income was used after deducting expenses. Farm household income was calculated as income from crop cultivation + income from livestock + income from regular wages and salaries + income from casual wages + income from non-farm business

4.3.3 Productivity

The enquiry into the farm-size and productivity relationship requires a specific approach. The relationship can be measured using approaches like each crop-wise analysis or farm-level analysis depending upon the objective to achieve. The difference in productivity on different farms can be due to the following factors:

1. Land quality of the farm (intrinsic property of land).
2. Price received by the farmers (marketing system).
3. Crop selection, cultivation practices including input-use (farm management).

To check the role of land quality variable on the productivity of different farm-size categories, it is best to control price and other varying factors. In practice, it is impossible to control other variables but their impact can be minimized by crop-wise analysis. The productivity is calculated as under:

A) Physical productivity of individual crops (Kg/ha)

B) Value productivity of a farm (Rs/ha)

A) Physical productivity of individual crops (Y_i)

$$Y_i = \frac{Q_i}{A_i}$$

Where, Y_i = Physical productivity of individual crops (Kg/ha)

Q_i = Production of crop under consideration for ith farm household (Kg)

A_i = Area under crop (ha)

B) Value productivity of a farm

The productivity of farms in different farm-size categories were calculated to analyse productivity differentials across different farm-size categories. Productivity for a farm can be affected by intrinsic properties, market conditions and farm management practices. It is well known that the intrinsic properties of the soil remain same in short-run. Further, the price received by the farmers may vary for different crops according to the quality of the produce or the market arrivals at the time of sale of produce. Similarly, the overall returns received by a farm may vary according to the choice of the crops grown in a year i.e., crop selection as well as cropping intensity. Hence, farm-level analysis will be a better approach to measure overall productivity of a farm. In farm-level analysis, the monetary value of all the crops grown on a farm in a specific period was used to analyze productivity differential among different farm-size categories. The productivity for a farm was determined at two levels:

a) Productivity for all crops (Rs/ha)

b) Productivity for kharif/rabi crops (Rs/ha)

a) Annual productivity: The overall productivity of farm i.e., value productivity was calculated at unit level by adding the total value of crops grown on a farm in an agricultural year. All the crops grown during the survey year was enlisted for each farm household. For each crop the total value of productivity was calculated as:

i) Gross value of individual crop (g_i)

$$g_i = q_i + p_i$$

Where, g i = Gross value of crop under consideration for ith farm household (Rs)

q i = Value of product (Rs)

p i = Value of by-product (Rs)

These individual values were summed-up to obtain overall crop value for a farm.

ii) Gross returns per annum (G_i)

$$G_i = \sum_{j=1}^{m} gij$$

Where, g ij = Gross value of jth crop for ith farm household (Rs)

m= number of crops on ith farm in a year

The value of productivity of all crops was then obtained by dividing the annual gross returns by the operational holdings in hectare.

iii) Productivity per hectare per annum (P_i ¬):

$$P_i = \frac{G_i}{(O.H.)i}$$

Where, G_i = Gross returns per annum for ith farm household (Rs.)

$(O.H)_i$ = Operational holding for ith farm household (ha)

b) Productivity for different seasons: The productivity per season for rabi and kharif crops was calculated by taking into account all the crops grown during the particular season by each farm household. At unit level, total number of crops grown per season was enlisted for each farm. For each individual crop, the monetary value was analysed by adding the value of product and by-product. Then, these individual values were summed-up to obtain overall value for a farm.

i) Gross returns per season (S_i)

$$S_i = \sum_{j=1}^{p} g_{ij}$$

Where, g ij = Gross value of jth crop for ith individual (Rs)

p= number of crops on ith farm in a season

The value of productivity of all crops grown in a season is then obtained by dividing the gross returns by area under the crop during the season.

ii) Productivity per hectare per season (Q_i¬):

$$Z_i = \frac{S_i}{(N.C.A)_i}$$

Where, S_i = Gross returns per season for ith farm hiusehold (Rs)

$(NCA)_i$ = cropped area for ith farm household in a given season (ha)

4.3.4 Paid-out cost (C_i)

The expenses incurred on crop production by the agricultural households were collected in the survey and unit-wise data was available. The inputs costs included the cost of seeds, fertilizers, manures, plant protection chemicals, human labour (hired), electricity, irrigation, diesel, repairs and maintenance of machinery, cost of hiring machinery, rent for leased-in land, interest on loans for cultivation purposes and other expenses. The sum of all the costs was termed as paid-out cost (C i)

4.3.5 Returns over paid-out cost per farm (N_i)

The returns over paid-out cost were calculated by subtracting the paid-out cost incurred on crop production from the total value of the produce on the ith farm.

$$N_i = G_i - C_i$$

Where, G_i = Gross returns per annum for ith farm household (Rs.)

C_i= Paid-out cost (Rs)

Per hectare returns over paid-out cost: It is calculated by dividing net returns by the operational holdings of each farm household.

$$R_i = \frac{N_i}{(O.H.)_i}$$

Where, N_i = Returns over paid-out cost

$(O.H)_i$ = Operational holding for ith farm household (ha)

4.4 ANALYTICAL PROCEDURES

The data was analyzed using various statistical techniques such as averages, percentages, Analysis of Variance (ANOVA), correlation, regression, Gini coefficient, Lorenz curve, Gini correlation and Theil index. Objective-wise details of the procedures and techniques used are given below:

4.4.1 Objective 1: To analyse farm size and productivity relationship

The data were critically scrutinized before considering the analytical techniques used to achieve the stated objectives.

A) Outlier detection and handling

Sometimes it may be found that the data contains extreme values which do not lie in the normal range. These observations are called outliers. Outliers in the data may arise due to sampling error or non-sampling error. Removal of outliers, where necessary, helps to remove erroneous observations in the sample. It is a healthy practice to obtain reliable statistics for the population. In present study, productivity was calculated and scatter plot was obtained with area under cultivation as x-axis and productivity as y-axis (see appendix 1). The data was inspected visually. Thereafter, Inter Quartile Range (IQR) approach was used to detect outliers (Vinutha et al 2018). Given the nature of farm level data, the range of Q_1 - 3*(IQR) and Q3 + 3*(IQR) was used to detect most probable outliers.

$$\text{Lower limit} = [Q_1 - (3 \times IQR)]$$

$$\text{Upper limit} = [Q_3 + (3 \times IQR)]$$

$$IQR = Q_3 - Q_1$$

Where, IQR is the value of difference between the value of quartile 3 and quartile 1.

Objective of this exercise was to remove extreme outliers which were observed during the scrutiny of the data. The observation with exceptionally extreme values for productivity were removed from the sample.

B) Analysis of variance

One-way ANOVA test is statistical technique used to test more than two independent groups. It is helpful to determine if there is statistical evidence that these groups have same population mean or different population mean. The test was applied to check the productivity difference across different farm-size categories for i) individual crop (Kg/ha), ii) kharif crops (Rs/ha), iii) rabi crops (Rs/ha) and iv) overall value productivity for an agricultural year (Rs/ha) among different farm-size categories.

$$F = \frac{MST}{MSE}$$

$$MST = \frac{\sum_{I=1}^{K}\left(\frac{T_i^2}{n_i}\right) - G^2/n}{k-1} \quad \text{and} \quad MSE = \frac{\sum_{I}^{K}\sum_{j=1}^{n_i} y_i^2 j - \sum_{i=1}^{k}\left(\frac{T_i^2}{n_i}\right)}{n-k}$$

Where, F = variance ratio for the overall test.

MST = mean square due to group (between group),

MSE = mean square due to error (within group),

Y_{ij} = observation,

T_{ij} = group total,

G = grand total of all observations,

Ni = number of observations

The obtained F-value for ANOVA is compared with the table value for f-statistics. If F¬cal > Ftab, the results are significant i.e., there is at least one group with mean significantly different from the other groups. We can apply post hoc test to further ascertain which group is statistically different.

C) Post-hoc test

Post-hoc test has been used to identify the pairs with significantly different means when the results for ANOVA are significant. Post-hoc test find difference between multiple groups while controlling the experiment-wise error rate.

Tukey-Kramer test: It is a post-hoc test applied on the independent groups. The Tukey-Kramer test explores which pair-wise comparison are significantly different.

$$q_s = \frac{\overline{X}_a - \overline{X}_b}{s.e.}$$

Where $\overline{X}_a$ is the mean of group with bigger mean out of the pair under consideration and $\overline{X}_b$ is mean for other group, in each pair-wise comparison.

$$s.e. = \sqrt{\frac{MSE}{2}\left(\frac{1}{n_i} + \frac{1}{n_j}\right)}$$

if qs value is larger than the critical value obtained from studentized range distribution (qα), the two means are said to be significantly different at level of significance α.

D) Correlation

It is statistical method that helps to measure the degree of association between two variables X and Y. In present study, the farm size and productivity were taken as X and Y.

$$r = \frac{n\left(\sum_{i=1}^{n} xy\right) - \sum_{i=1}^{n} x \sum_{i=1}^{n} y}{\sqrt{\left[n\sum_{i=1}^{n} x^2 - \left(\sum_{i=1}^{n} x\right)^2\right]\left[n\sum_{i}^{n} = 1y^2 - \left(\sum_{i=1}^{n} y\right)^2\right]}}$$

Where, r = correlation coefficient

x and y are the deviations from their respective means

The value of correlation coefficient (r) lies between -1 to 1. Positive value of correlation coefficient implies that both the variables move in the same direction i.e., with increase in one variable the other variable also increases. The negative value implies that the variables move in opposite direction i.e., with increase in one variable the other variable decreases. If the value of correlation coefficient is zero, it implies no linear relationship between variables. Farther the value of coefficient from zero, stronger the relation i.e., as we move away from zero toward one the relationship will be strongly positive and as we move away from zero toward minus one the relationship will be strongly negative.

E) Regression

It is a statistical method which helps us to study the relationship between two or more continuous variables. Out of all these variables, one is an outcome variable and other are explanatory variables. Outcome variable is commonly known as dependent variable while explanatory variables are also called as independent variables or predictor variables. Due to ease of computational software, different models are applied on the data to see better fit. Two models were fitted as under:

a) Simple linear regression model

b) Log-log regression model

a) Simple linear regression

When in a regression, there is single dependent variable and it is linear and not polynomial, the regression is termed as simple linear regression. In present study, the independent variables under consideration were operational holdings (ha), area under kharif crops (ha), area under rabi crops (ha) and area under wheat (ha) for different analysis. The dependent variables were productivity of wheat (Kg/ha), value productivity of kharif crops (Rs/ha), value productivity of rabi crops (Rs/ha), annual value productivity (Rs/ha), net returns (Rs/ha) and paid-out costs (Rs/ha) for different analysis.

$Y = a + bX$

Where, Y=dependent variable (Kg/ha or Rs/ha)

X=independent variable (Rs/ha)

a=intercept

b=slope

b) Log-log regression

The study conducted by Rudra (1968a) critically evaluated the exclusive use of linear regression in farm-size productivity studies. He suggested that other non-linear relationships should also be analyzed if there is no linear relationship between the farm-size and productivity. Therefore, log-log regression model was also considered for testing the farm-size and productivity relationship where linear model was rejected. In log-log model the data is transformed to log scaled for both the variables. The independent variables under consideration were the log values of operational holdings (ha), area under kharif crops (ha), area under rabi crops (ha) and area under wheat (ha) for different sections of analysis. The dependent variables were the log value of productivity of wheat (Kg/ha), value productivity of kharif crops (Rs/ha), value productivity of rabi crops (Rs/ha), annual value productivity (Rs/ha), net returns (Rs/ha) and paid-out costs (Rs/ha) for different sections of analysis.

$$\log Y = \beta_o + \beta_1 \log X$$

Where, Y = dependent variable

X = independent variable

β_o = intercept

β_1 = slope

The model with the higher coefficient of determination (R_2) was constituted as better fit and retained.

4.4.2 Objective 2: To estimate farm and non-farm income of farm household of different farm size categories

Simple statistical tools like averages and percentages were used to analyze income of a farm household and share of different sources in total income.

All the enlisted sources of incomes are added to obtain annual income of a farm household. To work-out total income of farm household:

Income from crops and livestock are estimated by deducting paid-out costs from gross returns.

Net farm business income was used after deducting expenses.

Farm household income = income from crop cultivation + income from livestock + income from regular wages and salaries + income from casual wages + income from non-farm business

4.4.3 Objective 3: To study income distribution among farmers

As income distribution is a statistical concept which explains how income is distributed among individuals or households in the society in which the farm households are classified based on the income of the household as top quintile, next quintile and so on till bottom quintile, the number of households in each category will tell us the distribution of income among farm households. There were few households in the present study with negative values of income. Following the methods suggested by Bellu and Liberati (2006), the negative values were given a small positive value (e). The small positive value (e) in the present study was 10^{-5}.

A) Lorenz curve

It is a graphical method to represent distribution of income with respect to distribution of population. The data for income and sample population was arranged presented in ascending order to obtain per cent of cumulative total for income and sample population. In the present study, the per cent population was taken on the x-axis and the income of the farm households was taken on y-axis. The Lorenz curves were constructed for the total income of the farm household as well as the income of the farm household from each income source.

B) Quintile groups

Quintile can be defined as a special form of quantile in which the series is divided into five equal parts. For present study, the data was differentiated into quintile groups according to the total income of the household. The share of various sources of income in each quintile group as well as per cent share of different sources within each quintile was also analyzed.

4.4.4 Objective 4: To analyze determinants of income inequality among farmers

Income inequality is how unevenly income is distributed throughout a population. The less equal the distribution, the higher income inequality. Decomposition of income inequality, both vertical as well as horizontal, based on income sources and farm-size categories was performed. Various tools used for determining income inequality and to decompose it into various horizontal and vertical components were:

A) Gini coefficient

It is the measure of degree of concentration of income distribution.

$$G\frac{2\mathrm{cov}\left[y,f(y)\right]}{\bar{y}}$$

where, G = Gini coefficient of total income

y = total income

f(y) = cumulative distribution of income

$\bar{y}$ = mean income of the sample

The value of Gini coefficient lies between 0 and 1. Higher value of the coefficients signifies high inequality. Gini coefficient has been used as it satisfes all the desirable properties of an inequality measure and also decomposable by income sources.

B) Theil index

Theil index the only measure which satisfies desirable properties of an inequality measure and decomposable into within and between categories (Cowell, 2005). The Theil inequality coefficient is defined as:

$$T = \frac{1}{n\overline{\overline{X}}}\sum_{i=1}^{n} x_i \log\left(\frac{x_i}{\overline{\overline{X}}}\right)$$

for the set of discrete income observations xi where i varies from 1 to n with mean $\overline{X}$. The popularity of Theil index lies in the fact that the inequality obtained by the index can be naturally decomposed into between-group and within-group components when the income observations are classified mutually exclusive set (j = 1 to g) which is commonly denoted as horizontal decomposition of inequality.

$$T = T_{BG} + T_{-WG}$$

Where

$$T_{BG} = \sum_{J=1}^{G} \frac{n_j \overline{X}_j}{n\overline{\overline{X}}} \log\left(\frac{\overline{X}_j}{\overline{\overline{X}}}\right)$$

$$T_{WG} = \sum_{J=1}^{G} \frac{n_j \overline{X}_j}{n\overline{\overline{X}}} T_j$$

Where j equals number of groups (farm categories in present case), n and Nj is total number of households and number of households in group j, respectively; $\overline{\mathbf{X}}_j$ is income of the household in group j, and is mean income of all households. The first (TBG)and second (TWG) terms in the relation represents between-group and within-group inequality, respectively.

where Tj (j = 1 to g) is the Theil inequality coefficient for the income observations in the jth group:

$$T_j = \frac{1}{n_j\overline{\overline{X}}_j}\sum_{i=1}^{nj} X_{ij} \log\left(\frac{X_{ij}}{\overline{\overline{X}}_j}\right)$$

Again, if all income observations within each group are equal, the within-group inequality disappears and the overall Theil coefficient reduces to the between group component TBG. It is important to note that the index does not give weights to the individual groups. Theil index was computed for all the income sources and total income of the household by differentiating the income by farm-size category groups.

C) Vertical decomposition of income inequality

There were five income sources in the present study i.e., income from cultivation, livestock income, income from salaries, income from casual wages and non-farm business. The contribution of each income source and it effect on total income distribution was obtained by the vertical decomposition of income inequality using method by Lerman and Yitzhaki (1985). It provide various coefficients Sk , Gk and Rk in which Sk represents the share of source k in total income and the coefficient obtained depicts the importance of income source with respect to total income; Gk is the source Gini which corresponds to the distribution of income from the source k and its coefficient quantifies the inequality caused by given income sources in total inequality; and Rk is the Gini correlation which gives the correlation between distribution of income from source k and the distribution of total income indicating how a given income source is correlated to the total income of a household.

$$G_k = 2 * CoV\left(\frac{Y_k F_k}{\overline{Y}_k}\right)$$

$$R_k = \frac{CoV(Y_k, F)}{CoV(Y_k, F_k)}$$

Where, $\overline{Y}_k$ the mean income from income source k, CoV (Yk , Fk) is the covariance between income component k and its cumulative distribution, CoV (Yk , F) is the covariance between income component k and cumulative distribution of the total income.

Further, using the Gini decomposition by income source, the effect of changes in a particular component on inequality can be estimated, holding income constant for all other sources. Assuming a change in each household's income from source k equal to ek, where e is close to 1, then the partial derivative of the Gini coefficient with respect to a percentage change e in source k will be:

$$\frac{aG}{ae_k} = S_k (G_k R_k - G)$$

Then, the marginal effect of the income source relative to the overall Gini can be obtained by dividing by overall Gini coefficient (G) as follows:

$$\frac{aG / ae}{G} = \frac{S_k G_k R_k}{G} S_k$$

Robustness of the marginal effect was examined by bootstrapping techniques following the methodology described by Kimhi et al (2006). It helped to derive reliable stand errors for the statistics obtained.

Chapter-5

Scio-Economic Profile of the Farm Households in Punjab

This chapter helps to understand the socio-economic profile of the farm households in Punjab. Major objective of this section was to report reporting some preliminary information about the basic characteristics of the household so that broader idea about their overall status can be drawn.

5.1 FAMILY SIZE OF THE FARM HOUSEHOLDS

The average family size and number of the respondent households across different farm size categories has been presented in the Table 5.1. The data pertains to total 558 farm households. There were 108 (19.35%), 142 (25.44%), 167 (29.92%), 120 (17.92%) and 21 (3.76%) marginal, small, semi-medium, medium and large farm households, respectively. The average family size for Punjab was 5.49 members per households. The average family size was 4.81, 5.05, 5.46, 6.38 and 7.05 members for marginal, small, semi-medium, medium and large farm households, respectively indicating direct relationship with farm size.

Table 5.1: Family size of the farm households across different farm-size categories, Punjab

Particulars	**Marginal** (<1.00 ha)	**Small** (1.01-2.00 ha)	**Semi-medium** (2.01-4.00 ha)	**Medium** (4.01-10.00 ha)	**Large** (>10.00 ha)	**Overall**
Number of households	108 (19.35)	142 (25.44)	167 (29.92)	120 (17.92)	21 (3.76)	558 (100.00)
Average family size	4.81	5.05	5.46	6.38	7.05	5.49

5.2 COMPOSITION OF OPERATIONAL HOLDINGS FOR DIFFERENT FARM-SIZE CATEGORIES

The operational holding of a household is the sum of owned & managed land and leased-in land. The owned & managed land is obtained after deducting the leased-out land out of owned land. The information on the average size of operational holding for different farm-size categories has been presented in Table 5.2. The average size of operational holdings in Punjab was 3.12 hectares. The per cent share of owned & managed land in total operational holding was 74.28 per cent. The share of lease-in land was 25.72 per cent. The average operational holdings across different farm-size categories were 0.47, 1.40, 2.76, 6.00 and 14.81 hectares for marginal, small, semi-medium, medium and large farms, respectively. The share of leased-in land in the operational holding increased with increase in farm size. It was 8.76, 13.25, 20.65, 28.33 and 37.93 per cent for marginal, small, semi-medium, medium and large farm-size categories, respectively.

Table 5.2: Operational holdings of the farm households, Punjab

Particulars	Marginal (n_1=108)	Small (n_2=142)	Semi-medium (n_3=167)	Medium (n_4=120)	Large (n_5=21)	Overall (n=558)
a) Owned land	67.17	192.67	396.62	531.78	204.38	1392.62
b) Leased-out	20.4	20.72	31.03	15.97	11.33	99.45
c) Owned and managed (a+b)	46.77 (91.24)	171.95 (86.75)	365.59 (79.35)	515.81 (71.67)	193.05 (62.07)	1293.17 (74.28)
d) Leased-in	4.49 (8.76)	26.26 (13.25)	95.14 (20.65)	203.93 (28.33)	117.97 (37.93)	447.79 (25.72)
e) Operational holdings	**51.26 (100.00)**	**198.21 (100.00)**	**460.73 (100.00)**	**719.74 (100.00)**	**311.02 (100.00)**	**1740.96 (100.00)**
Average operational holdings (ha)	**0.47**	**1.40**	**2.76**	**6.00**	**14.81**	**3.12**

Note: Figures in the parentheses are percentage to their respective total operational holdings

5.3 DISTRIBUTION OF FARM HOUSEHOLDS ON THE BASIS OF CASTE

In India, the society is divided into number of social groups and Punjab is no exception to it. The information on the caste of farm households in Punjab is given in Table 5.3. Most of the farm households in Punjab belonged to general category (84.41%) followed by Other Backward Class (OBC) (9.32%) and Schedule Caste (SC)/Schedule Tribe (ST) class (6.27%). In different farm-size categories, the distribution among caste was uneven although predominant share was available to general caste in each farm-size category. The general category constituted more

Chapter-6

Cropping Pattern, Cost and Returns Structure Among Farm Households

The cropping pattern, cost and returns structure of cost of cultivation have been presented in this section. The per hectare cost and returns structure has been also analyzed. The section has been divided into following subsections:

6.1 Cropping pattern among farm households

6.2 Gross returns in crop cultivation among farm households

6.3 Paid-out cost structure among farm households

6.4 Crop returns over paid-out cost among farm households

6.1 CROPPING PATTERN AMONG FARM HOUSEHOLDS

The cropping pattern among the farm households in Punjab for the survey year has been presented in the Table 6.1. The sample size for the Punjab was 558 farm households. Per annum value of output per hectare was computed. The data was scrutinized for potential outliers. The outliers were detected and removed. The final sample retained was 539 farm households with 102 (18.92%) marginal households, 133 (24.68%) small households, 163 (30.24%) semi-medium households, 120 (22.26%) medium households and 21 (3.90%) large households. It was observed that wheat has the highest area under cultivation (41.12%) followed by paddy (35.34%). Both these crops comprised 76.46 per cent of total cropped area during the survey year. Among other crops, fodder has major share (9.59%), followed by cotton (4.93%), maize (2.35%), potato (2.19%) and sugarcane (1.25%). Other miscellaneous crops occupied 3.23 per cent share in the total cultivation.

Table 6.1: Cropping pattern across different farm-size categories, Punjab

Particulars	Marginal	Small	Semi-medium	Medium	Large	Overall
Wheat	27.36 (30.98)	138.53 (41.53)	358.24 (42.91)	580.89 (42.76)	221.74 (36.25)	1326.76 (41.12)
Paddy	19.50 (22.08)	110.40 (33.10)	296.07 (35.47)	520.01 (38.28)	194.34 (31.77)	1140.32 (35.34)
Cotton	3.93 (4.45)	22.25 (6.67)	29.47 (3.53)	61.08 (4.50)	42.4 (6.93)	159.13 (4.93)
Maize	1.77 (2.00)	6.25 (1.87)	24.18 (2.90)	26.51 (1.95)	17.00 (2.78)	75.71 (2.35)
Potato	-	-	-	18.61 (1.37)	51.95 (8.49)	70.56 (2.19)
Sugarcane	-	0.25 (0.07)	5.10 (0.61)	12.44 (0.92)	22.46 (3.67)	40.25 (1.25)
Fodder	30.65 (34.71)	43.22 (12.96)	91.82 (11.00)	106.28 (7.82)	37.58 (6.14)	309.55 (9.59)
Miscellaneous crops	5.10 (5.78)	12.63 (3.79)	29.93 (3.59)	32.52 (2.39)	24.15 (3.95)	104.33 (3.23)
GCA	**88.31 (100.00)**	**333.53 (100.00)**	**834.81 (100.00)**	**1358.34 (100.00)**	**611.62 (100.00)**	**3226.61 (100.00)**
CI (%)	**179.46**	**179.71**	**185.75**	**188.73**	**196.65**	**188.14**
Correlation coefficient	**0.15*****					

GCA=Gross Cropped Area; CI=Cropping Intensity
***significant at 1% level of significance
Note: Figures in the parentheses are percentage to their respective GCA

The share of different crops across farm-size categories has significant variation. In marginal farm-size category, the fodder crop has the highest share (34.7%) in GCA. It was followed by wheat (30.98%) and paddy (22.08%). For other farm-size categories, wheat and paddy dominated the area under cultivation. It is interesting to note that on large farms, along with wheat (36.25%) and paddy (31.77%), other cereals like maize (2.78%) and other cash crops like cotton (6.93%), potato (8.49%) and sugarcane (3.67%) were also grown extensively. It is well known that these crops are highly capital intensive. More or less similar pattern was observed on medium farms also. The overall cropping intensity for Punjab among the farm households was found to be 188.14 per cent. Across different farm-size, there was upward trend in cropping intensity as the farm-size increased. The cropping intensity for marginal, small, semi-medium, medium and large farms was 179.46, 179.71, 185.75, 188.73 and 196.65 per cent, respectively. Correlation coefficient between cropping intensity and farm-size slightly positive (r=0.15) and significant (p<0.001). In earlier studies, the relationship between cropping intensity and farm-size categories was found to be negative for Punjab agriculture i.e., small farmers were cultivating the land more intensively (Rudra and Chattopadhyay 1976, Bhalla and Chadda 1982).

6.2 GROSS RETURNS IN CROP CULTIVATION AMONG FARM HOUSEHOLDS

The gross returns from different crops cultivated by the farm households during survey year have been presented in the Table 6.2. The overall value of product, value of by-product and gross returns per annum for farms in Punjab were Rs 4.24 lakh, Rs 25,550 and Rs 4.50 lakh, respectively.

Table 6.2: Gross returns in cost of cultivation across different farm-size categories, Punjab

Particulars	Marginal	Small	Semi-medium	Medium	Large	Overall
per farm						
Value of product	54852 (93.40)	166750 (92.44)	343793 (92.39)	818148 (95.04)	2216490 (96.23)	423998 (94.32)
Value of by-product	3873 (6.60)	13639 (7.56)	28320 (7.61)	42704 (4.96)	86746 (3.77)	25550 (5.68)
Total value	**58725 (100.00)**	**180389 (100.00)**	**372113 (100.00)**	**860852 (100.00)**	**2303237 (100.00)**	**449548 (100.00)**
per ha						
Value of product	113717 (93.40)	119486 (92.44)	124690 (92.39)	136409 (95.04)	149657 (96.23)	133258 (94.32)
Value of by-product	8029 (6.60)	9773 (7.56)	10271 (7.61)	7120 (4.96)	5857 (3.77)	8030 (5.68)
Total value	**121747 (100.00)**	**129259 (100.00)**	**134962 (100.00)**	**143529 (100.00)**	**155514 (100.00)**	**141288 (100.00)**

Note: Figures in the parentheses are percentage to the total value.

The value of product per farm and the value of by-product per farm increased with increase in farm-size. The gross returns per farm was Rs 0.59 lakh, Rs 1.80 lakh, Rs 3.72 lakh, Rs 8.61 lakh and Rs 23.03 lakh for marginal, small, semi-medium, medium and large farm-size categories, respectively. The gross returns per hectare across different farm-size categories were also analyzed to examine the productivity of different farms. The overall gross returns per hectare for all farm-size categories was Rs 1.41 lakh. Gross returns per hectare for the large farms was the highest (Rs 1.56 lakh), followed by medium farms (Rs 1.43 lakh), semi-medium farms (Rs 1.35 lakh), small farms (Rs 1.29 lakh) and the lowest (Rs 1.22 lakh) for marginal farms.

6.3 PAID-OUT COST STRUCTURE IN CROP CULTIVATION AMONG FARM HOUSEHOLDS

The annual paid-out cost on crop cultivation includes cost of seeds, fertilizers, manures, plant protection chemicals, human labour, electricity, irrigation, diesel, repairs and maintenance, cost of hiring machinery, rent for leased-in land, interest on loans for cultivation purposes and other expenses. The paid-out costs for different farm-size categories for Punjab were presented in Table 6.3. Overall total cost per farm was Rs 1.98 lakh out of which the highest share was Rs 52,061 (26.32%) for rent of leased-in land. It was followed by expenses on fertilizer and manures which was Rs 30,192 (15.27%) and human labour which was Rs 27,836 (14.07%). The next important expenses were diesel (11.43%), plant protection chemicals (10.40%) and cost of hiring machinery (7.73%). Other important costs were seed (5.42%) and minor repairs (3.20%). Other minor expenses and interest on loans for cultivation comprises of 6.16 per cent of the total paid-out cost. On marginal farms, share of fertilizer and manures was the highest (18.42%) followed by hiring machinery (14.87%), rent for leased-in land (13.50%), human labour (11.74%), seeds (10.74%), plant protection chemicals (9.75%), etc. More or less same pattern was observed on small farms with the highest share of fertilizer and manures (18.00%) followed by human labour (13.90%), plant protection chemicals (12.26%), diesel (12.04%), rent for leased-in land (11.75%), hiring machinery (11.60%), etc. On semi-medium farms, the highest share in cost was for rent for leased-in land (19.61%), followed by fertilizer and manures (16.26%), human labour (15.01%), diesel (12.69%), plant protection chemicals (10.29%), hiring machinery (8.99%), seeds (6.09%), etc. The hierarchy of cost share on medium farms was similar to semi-medium farms with only difference of higher per cent share in rent for leased-in land (29.02%). It was followed by fertilizer and manures (14.73%), human labour (12.98%), diesel (12.42%), plant protection chemicals (10.54%), hiring machinery (7.00%), seeds (4.86%), etc. On large farms, rented land has the highest share in total paid-out cost (33.76%). Human labour was the second highest cost (15.35%) followed by fertilizer and manures (14.10%), plant protection chemicals (9.65%), diesel (8.60%), hiring machinery (5.95%), seeds (4.66%), etc.

The details on paid-out costs per hectare across different farm-size categories have been presented in Table 6.4. It helped to analyze the pattern of expenditure of different farm-size categories on different cost items. The rent for leased-in land was the lowest on small farms (Rs 5696) whereas it was the highest on larger farms (Rs 28579). It clearly depicted that more land was leased-in by large farm-size categories. The use of fertilizers also followed the same pattern. The per hectare cost on fertilizers and manure was the lowest on marginal farms (Rs 7869) whereas it was the highest on larger farms (Rs 11933). The trend was upward for human labour, diesel and plant protection chemicals also. The cost incurred on human labour per hectare was the lowest on marginal farms (Rs 5014) and the highest on large farms (Rs 12995). The value of diesel per hectare was the lowest on marginal farms (Rs 3771) and the highest on medium farms (Rs 7821). The cost of plant protection chemicals per hectare was the lowest on marginal farms (Rs 4165) and the highest on large farms (Rs 8173). The cost of hiring machinery followed opposite trend. It was the highest on marginal farms (Rs 6352) and the lowest on medium farms (Rs 4411). The cost of seeds per hectare also followed the same pattern as hired machinery. It was the highest on marginal farms (Rs 4587) and the lowest on medium farms (Rs 3061). Expenses per hectare incurred on minor repairs were highest on semi-medium farms. Interest on loans for cultivation was the lowest on marginal farms (Rs 1477) and the highest on large farms (Rs 4175).

Table 6.3: Paid-out cost structure across different farm-size categories, Punjab

Paid-out Costs	Marginal	Small	Semi-medium	Medium	Large	Overall
Rent for leased-in land	2782 (13.50)	7949 (11.75)	28680 (19.61)	109640 (29.02)	423262 (33.76)	52061 (26.32)
Fertilizer and manure	3796 (18.42)	12175 (18.00)	23777 (16.26)	55667 (14.73)	176729 (14.10)	30192 (15.27)
Human labour	2419 (11.74)	9405 (13.90)	21948 (15.01)	49058 (12.98)	192457 (15.35)	27836 (14.07)
Diesel	1819 (8.83)	8147 (12.04)	18560 (12.69)	46906 (12.42)	107819 (8.60)	22611 (11.43)
Plant protection chemicals	2009 (9.75)	8291 (12.26)	15045 (10.29)	39831 (10.54)	121040 (9.65)	20560 (10.40)
Cost of hired machinery	3064 (14.87)	7847 (11.60)	13140 (8.99)	26459 (7.00)	74555 (5.95)	15285 (7.73)
Seeds	2213 (10.74)	5014 (7.41)	8909 (6.09)	18359 (4.86)	58385 (4.66)	10712 (5.42)
Minor repairs and maintenance	759 (3.68)	2994 (4.43)	7442 (5.09)	11719 (3.10)	15180 (1.21)	6334 (3.20)
Irrigation	412 (2.00)	667 (0.99)	263 (0.18)	308 (0.08)	8095 (0.65)	706 (0.36)
Other expenses	624 (3.03)	1717 (2.54)	3496 (2.39)	6747 (1.79)	14462 (1.15)	3665 (1.85)
Interest on loans for cultivation	713 (3.46)	3440 (5.09)	4977 (3.40)	13115 (3.47)	61837 (4.93)	7818 (3.95)
Total paid-out cost	**20609 (100.00)**	**67646 (100.00)**	**146238 (100.00)**	**377809 (100.00)**	**1253822 (100.00)**	**197779 (100.00)**

Note: Figures in the parentheses are percentage to the total paid out cost.

Table 6.4: Paid-out costs per hectare across different farm-size categories, Punjab

(Rs/hectare)

Paid-out Costs	Marginal	Small	Semi-medium	Medium	Large	Overall
Rent for leased-in land	5767 (13.50)	5696 (11.75)	10402 (19.61)	18280 (29.02)	28579 (33.76)	16362 (26.32)
Fertilizer and manure	7869 (18.42)	8724 (18.00)	8624 (16.26)	9281 (14.73)	11933 (14.10)	9489 (15.27)
Human labour	5014 (11.74)	6739 (13.90)	7960 (15.01)	8179 (12.98)	12995 (15.35)	8749 (14.07)
Diesel	3771 (8.83)	5838 (12.04)	6732 (12.69)	7821 (12.42)	7280 (8.60)	7106 (11.43)
Plant protection chemicals	4165 (9.75)	5941 (12.26)	5457 (10.29)	6641 (10.54)	8173 (9.65)	6462 (10.40)
Cost of hired machinery	6352 (14.87)	5623 (11.60)	4766 (8.99)	4411 (7.00)	5034 (5.95)	4804 (7.73)
Seeds	4587 (10.74)	3593 (7.41)	3231 (6.09)	3061 (4.86)	3942 (4.66)	3367 (5.42)
Minor repairs and maintenance	1574 (3.68)	2146 (4.43)	2699 (5.09)	1954 (3.10)	1025 (1.21)	1991 (3.20)
Irrigation	854 (2.00)	478 (0.99)	95 (0.18)	51 (0.08)	547 (0.65)	222 (0.36)
Other expenses	1293 (3.03)	1231 (2.54)	1267 (2.39)	1125 (1.79)	976 (1.15)	1151 (1.85)
Interest on loans for cultivation	1477 (3.46)	2465 (5.09)	1805 (3.40)	2187 (3.47)	4175 (4.93)	2457 (3.95)
Total paid-out cost	**42725 (100.00)**	**48472 (100.00)**	**53039 (100.00)**	**62992 (100.00)**	**84658 (100.00)**	**62160 (100.00)**

Note: Figures in the parentheses are percentage to the total value.

The pattern of different paid-out costs across different farm-size categories have been presented in figures below. The Figure 1a, Figure 1b, Figure 1c and Figure 1d depicted the trend for paid-out costs per hectare for different inputs use. The Figure 1a presented paid-out cost incurred on rent for leased-in land. It can be observed that the expenditure on rent for leased-in land per hectare followed upward trend. There was very steep rise in leased-in land for large farmers. The Figure 1b presented paid-out cost incurred on human labour. The paid-out cost on human labour also showed upward trend. The Figure 1c depicted per hectare cost on seeds, plant protection chemicals and fertilizers & manure. The cost of seeds on per hectare basis showed declining trend with increase in farm-size. Plant protection chemicals and fertilizers & manure were positively related with farm-size. The Figure 1d contains per hectare cost on diesel, hired machinery and repair & maintenance. The cost of diesel for each hectare showed upward trend while per hectare cost of hiring machinery per hectare showed inverse trend i.e., higher expenses per hectare for marginal farms and lower expenses per hectare for medium and large farms. Minor repairs followed inverted U-shape trend as it was more on semi-medium and medium farms while less on marginal and large farm-size categories.

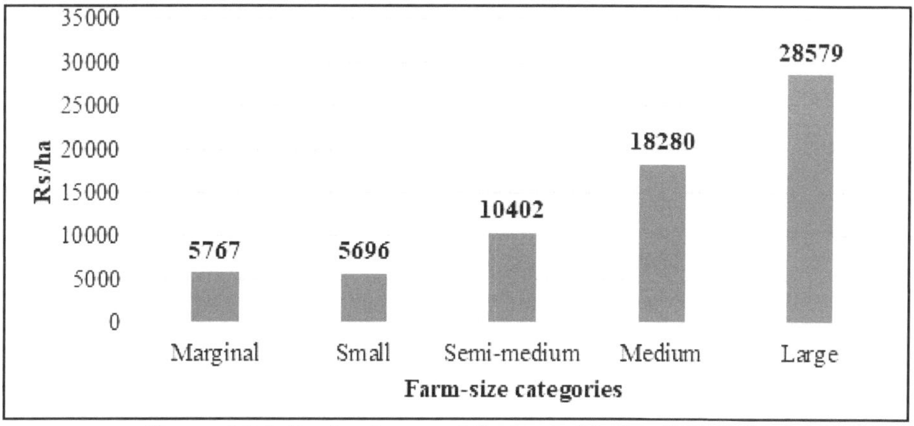

Figure 1 (a): Per hectare rent for leased-in land, Punjab

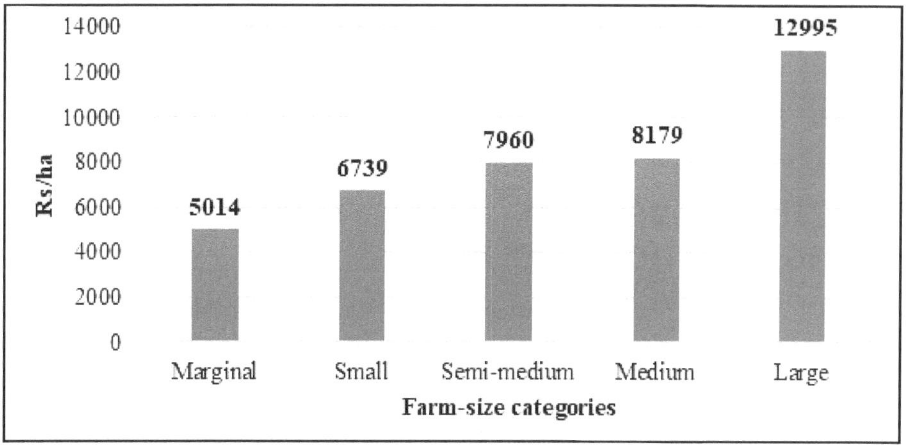

Figure 1 (b): Per hectare expenses on hired human labour, Punjab

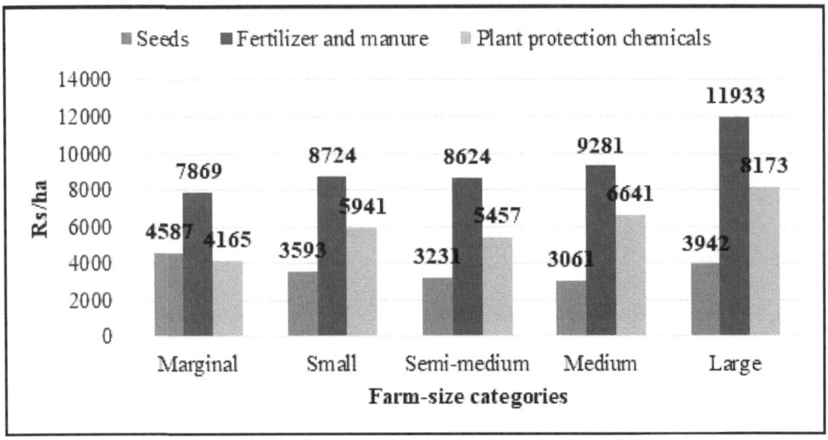

Figure 1(c): Per hectare expenses on seeds, plant protection chemicals and fertilizer & manures, Punjab

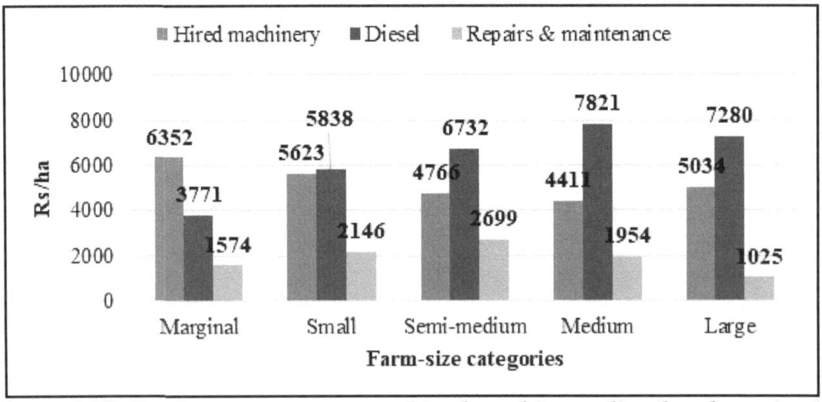

Figure 1(d): Per hectare expenses on hired machinery, diesel and repairs & maintenance, Punjab

6.4 RETURNS OVER PAID-OUT COST IN CROP CULTIVATION AMONG FARM HOUSEHOLDS

Gross returns, paid-out costs and net returns of crop cultivation across different farm-size categories has been presented in the Table 6.5. The value of net returns over paid-out cost was analyzed across different farm-size categories by deducting the paid-out cost from gross returns. The gross returns per farm, paid-out costs per farm and net returns over paid-out cost increased with increase in farm-size. The overall value for gross returns, paid-out costs and returns over paid-out cost was Rs 4.49 lakh, Rs 1.98 lakh and Rs 2.52 lakh, respectively. The value of returns over paid-out cost was Rs 38116, Rs 1.12 lakh, Rs 2.26 lakh, Rs 4.83 lakh and 10.49 lakh for marginal, small, semi-medium, medium and large farm-size categories, respectively. Similarly, per capita net returns were analyzed by deducting the per capita paid-out cost from per capita gross returns. The overall value for per capita gross returns, paid-out costs and returns over paid-out cost was Rs 81,585, Rs 35,893 and Rs 45,691, respectively. The value of per capita returns over paid-out cost was Rs 7870, Rs 22,347, Rs 41,322, Rs 75,672 and Rs 1,48,904 for marginal, small, semi-medium, medium and large farm-size categories, respectively. In per hectare terms, the value of returns over paid-out cost had inverted U shape. It increased up to semi-medium farms and then showed declining trend. It was Rs 79,022, Rs 80,786, Rs 81,923, Rs 80,537 and Rs 70,856 for marginal, small, semi-medium, medium and large farm-size categories, respectively. It was very interesting observation that in term of gross returns the large farmers earned the highest gross returns (Rs 1.56 lakh) but returns over paid-out cost were the lowest among all the farm-size categories (Rs 70,856). It was mainly because of high land rent and high expenditure on labour and machinery use.

Table 6.5: Gross returns, paid-out costs and returns over paid-out cost in cost of cultivation across different farm-size categories, Punjab

(Rs/annum)

Particulars	Marginal	Small	Semi-medium	Medium	Large	Overall
per farm						
Gross returns (a)	58725 (100.00)	180389 (100.00)	372113 (100.00)	860852 (100.00)	2303237 (100.00)	449548 (100.00)
Paid-out cost (b)	20609 (35.09)	67646 (37.50)	146238 (39.30)	377809 (43.89)	1253822 (54.44)	197779 (44.00)
Returns over paid-out cost (a-b)	38116 (64.91)	112743 (62.50)	225876 (60.70)	483042 (56.11)	1049415 (45.56)	251768 (56.00)
per capita						
Gross returns (a)	12125 (100.00)	35755 (100.00)	68075 (100.00)	134859 (100.00)	326811 (100.00)	81585 (100.00)
Paid-out cost (b)	4255 (35.09)	13408 (37.50)	26753 (39.30)	59187 (43.89)	177907 (54.44)	35893 (44.00)
Returns over paid-out cost (a-b)	7870 (64.91)	22347 (62.50)	41322 (60.70)	75672 (56.11)	148904 (45.56)	45691 (56.00)
per ha						
Gross returns (a)	121747 (100.00)	129259 (100.00)	134962 (100.00)	143529 (100.00)	155514 (100.00)	141288 (100.00)
Paid-out cost (b)	42725 (35.09)	48472 (37.50)	53039 (39.30)	62992 (43.89)	84658 (54.44)	62160 (44.00)
Returns over paid-out cost (a-b)	**79022 (64.91)**	**80786 (62.50)**	**81923 (60.70)**	**80537 (56.11)**	**70856 (45.56)**	**79128 (56.00)**

Note: Figures in the parentheses are percentage to gross returns.

To conclude, the main crops in among farm household in Punjab were wheat and paddy with around 76 per cent area under these crops. Other important crops were cotton (4.93%), maize (2.35%), potato (2.19%) and sugarcane (1.25%). It was found that large farms grow relatively more cash crops like sugarcane, potato, cotton as compared to other farm-size categories while marginal category relied more on fodder crops. The cropping intensity had positive relationship with farm-size categories. The cropping intensity was 179.46, 179.71, 185.75, 188.73 and 196.65 per cent for marginal, small, semi-medium, medium and large farm-size categories, respectively. The annual gross returns per farm for farm household were 4.49 lakh. The gross returns were Rs 0.59 lakh, 1.80 lakh, Rs 3.72 lakh, Rs 8.61 lakh and Rs 23.03 lakh for marginal, small, semi-medium, medium and large farm-size categories, respectively. Structure of the paid-out cost revealed that rent for leased-in land has the highest share (26.32%) in paid-out cost, followed by fertilizers & manure (15.27%), human labour (14.07). The value of per capita returns over paid-out cost were Rs 7870 and Rs 22347, for marginal and small, respectively. In per hectare terms, the value of returns over paid-out cost had inverted U-shape. The large farmers earned the highest gross returns (Rs 1.56 lakh) but the returns over paid-out cost were the lowest among all the farm-size categories (Rs 70856).

Chapter-7

Farm-Size and Productivity Relationship in Punjab

Analysis of farm-size and productivity relationship enquires if there is difference in productivity across different farm-sizes. The IR between farm-size and productivity was reported in India (Sen 1962, Khusro, 1964, Mazumdar, 1965); but generalization for whole country was refused by providing empirical evidence for certain regions (Rao 1967, Bhattacharya and Saini 1972, Bharadwaj 1974, Chattopadhyay and Rudra 1976, Rudra and Sen 1980). For Punjab agriculture, it was reported that the relationship has been changed from negative to neutral till the end of twentieth century (Singh et al 2000). Present study will enquire about the relationship between farm size and productivity for Punjab in recent times. The analysis can be broadly conducted at two levels i.e., crop-wise analysis and farm level analysis. Thus, this section has been classified into following subsections:

7.1. Crop-wise analysis of farm-size and productivity relationship

7.2. Value productivity analysis of farm-size and productivity relationship

7.3. Relationship between input-use and farm-size

Each sub-section has included outlier detection followed by the preliminary testing of the data. The information on the farm-size and productivity relationship has been established using correlation coefficient. It assisted in determining the productivity trend across different farm-size categories. Thereafter, ANOVA along with post-hoc test was applied. ANOVA analysis helped us to determine if there was significant productivity difference across different farm-size categories. Further, post-hoc test helped to identify the significantly different groups. The information provided by ANOVA and post-hoc test can be considered as preliminary analysis of the data. Rudra (1977) found that the high or low productivity tendencies may operate for certain range of farm-size categories only. Therefore, further information on the farm-size and productivity relationship was established using intra-group correlation coefficients. Regression analysis was also conducted to measure the magnitude of the farm-size and productivity relationship and to determine the goodness of fit.

7.1. CROP-WISE ANALYSIS OF FARM-SIZE AND PRODUCTIVITY

The productivity of a farm depends on large number of factors which can be broadly classified into three groups i) land quality (intrinsic properties) ii) price received (marketing) iii) farm management practices. To check the role of land quality on productivity of different farms, the role of price and other variable factors should be controlled. Empirically, it is impossible to control all other factors but their impact can be minimised using crop-level analysis. It facilitates the comparison of physical productivity across different farms which is not possible at farm-level analysis as most of the farmers grow more than one crop. Hence, crop-wise analysis was conducted to see farm-size and productivity relationship in Punjab. Paddy and wheat were the main crops of kharif and rabi season in Punjab, respectively. Paddy crop has many inter-regional differences within Punjab in terms of varieties sown and price received. The difference between the productivity and price of basmati and parmal paddy was a major concern. Also, there were certain regions sowing high yielding varieties like PUSA 44 which has around one and half times higher productivity than other cultivars prevalent in the region. These features violate the objective of bringing homogeneous crop conditions required to conducting crop-wise analysis. Thus, the crop-wise analysis of the relationship was conducted for wheat only.

7.1.1 Production and productivity of wheat among farm households

The wheat crop is the most important crop of Punjab with around 3.6 million-hectare area under cultivation of wheat (Anonymous, 2019). The production and productivity of wheat crop among the farm households was given in Table 7.1. The analysis for productivity of wheat crop was carried out to examine the difference in productivity of the crops across different farm-size groups. It is an exercise to analyze the existence of productivity relationship with farm-size under homogeneous crop. Productivity of the wheat crop for each farm household was obtained by dividing the production of wheat on the farm by the area under the crop for the farm. There were 499 farm households sowing wheat crop. By examining the data for productivity, outliers were detected. Again, using Q1 - 3*(IQR) and Q3 + 3*(IQR), outliers were removed as it was done earlier. The final sample had 484 farmers sowing wheat crop, out of which 65 farmers were marginal, 128 farmers were small, 155 farmers were semi-medium, 115 farmers were medium category and 21 were large farmers. The production of wheat crop per farm was 1.67, 4.62, 9.84, 22.34 and 48.98 tonnes for marginal, small, semi-medium, medium and large farm-size categories, respectively. The overall productivity of wheat among the farm households was 4.44 tonnes per hectare. In different farm-size categories, the productivity per hectare was 4.1, 4.3, 4.3, 4.5 and 4.6 tonnes for marginal, small, semi-medium, medium and large, respectively. Overall correlation coefficient for farm-size and productivity was found to be positive (r=0.17) and significant but weak (Table 7.2).

Table 7.1: Production and productivity of wheat across different farm-size categories, Punjab

Particulars	**Marginal**	**Small**	**Semi-medium**	**Medium**	**Large**	**Overall**
No. of farmers	65 (13.43)	128 (26.45)	155 (32.02)	115 (23.76)	21 (4.34)	484 (100.00)
Total area under wheat cultivation (ha)	26.55	137.77	351.97	572.90	221.74	1310.93
Average area under wheat cultivation (ha)	0.41	1.08	2.27	4.98	10.56	2.71
Production (Tonnes/farm)	1.67	4.62	9.84	22.34	48.98	12.03
Productivity (Tonnes/ha)	**4.09**[a]	**4.29**[a]	**4.34**[a]	**4.48**[a]	**4.64**[a]	**4.44**

Note: Figures in the parentheses are percentage to the overall.

[a] no significant difference among categories with same alphabet.

Table 7.2: Correlation and regression coefficients for farm-size and wheat productivity, Punjab

Particulars	**Marginal**	**Small**	**Semi-medium**	**Medium**	**Large**	**Overall**
Correlation						
Coefficient	-0.081	0.157	0.131	0.304	-0.003	0.174
p-value	0.522	0.076	0.104	0.001	0.990	0.001
Regression (Overall)						
Coefficient (B_1)	54.32					
R^2	0.030					
p-value	0.001					

Further, ANOVA revealed that the wheat productivity of the farmers across different farm-size categories was statistically similar. It implies that although there was positive relationship between farm-size and productivity of wheat, but the difference in productivity was not high enough to categorize farm-size into different productivity groups. Moving one step further, intra-category correlation between farm-size and productivity for each farm-size category was worked-out. It was found that the correlation coefficients for small and semi-medium were positive and significant while non-significant for all other categories. Rudra (1977) mentioned of this phenomenon that the high or low productivity tendencies may operate for certain range of farm-size. Regression equation was also fitted using area under wheat cultivation (ha) as independent variable and productivity of wheat crop (Kg/ha) as dependent variable. It was observed that the both intercept and coefficient were significant and positive. It was found that one hectare increase in crop may lead to 54 kg increase in productivity. Overall equation was significant but the R-square was very low which signifies that productivity was a significant factor explaining wheat productivity but there exist more important factors other than farm-size which determines wheat productivity. To conclude, it is confirmed that there was no significant inverse relationship between farm-size and productivity existed in wheat crop in Punjab. The correlation coefficient for farm-size and productivity was neutral for marginal farms (negative but non-significant), positive for small farms, semi-medium and medium farms and neutral for large farms (negative but non-significant). It is concluded that though there is productivity difference among the wheat crop but it was not very pronounced whereas earlier studies have reported neutral relationship between farm-size and productivity for wheat crop in Punjab (Bhalla and Chadha 1982, Singh et al 2000).

7.2 VALUE PRODUCTIVITY ANALYSIS FOR FARM-SIZE AND PRODUCTIVITY RELATIONSHIP

The overall value productivity of a farms was calculated to analyze productivity differential among different farms. Productivity of a farm depends on large number of factors like land quality, crops grown, cropping intensity, price received, farm management practices, input use, etc. Farm level analysis incorporates all these factors. In present study the farm level analysis was conducted at three levels:

7.2.1 Farm-size and productivity relationship - Kharif crops

7.2.2 Farm-size and productivity relationship - Rabi crops

7.2.3 Farm-size and productivity relationship - All crops

7.21. Farm-size and productivity relationship-kharif crops

During kharif season, paddy was the main in Punjab. There are other kharif crops like cotton, maize, etc., prevalent in the region. Their productivity cannot be compared in physical terms. Thus, the analysis was carried-out in value terms. There were 553 observations for kharif season. Outliers were removed using interquartile range for value productivity. The final sample had 539 farmers sowing kharif crops, out of which 99 farmers belonged to marginal category, 136 farmers were small category, 163 farmers were semi-medium, 120 farmers were medium category and 21 were large farmers. The total value of the crop per farm on the marginal farms was Rs 29927 and on the large farms was Rs 12.9 lakh (Table 4.3.3). The overall value of the kharif crops for all farm-size categories was Rs 2.49 lakh. Across different farm-size categories, the value productivity of kharif crops was the lowest (Rs 67490) in case of marginal farmers and the highest (Rs 84215) for medium farmers. Using correlation analysis, it was found that the relationship between productivity and cropped area during kharif season was significant and positive (r=0.15) but weak (Table 7.3). Although there was some difference in productivity on different farms but ANOVA analysis revealed that the productivity on all farm-size categories was statistically identical except marginal farms. Hence, the productivity of kharif crops can be broadly categorized into two groups; one was marginal farms where productivity was lower and other group constituted small, semi-medium, medium and large farm-size categories where productivity was higher. The intra-category correlation for each farm-size category was also analyzed.

Table 7.3: Gross returns from kharif crops across different farm-size categories, Punjab

(in Rs.)

Particulars	Marginal	Small	Semi-medium	Medium	Large	Overall
Rs/farm						
Value of product	28133 (94.00)	94992 (93.82)	185219 (94.62)	477870 (98.25)	1275487 (98.82)	241233 (96.95)
Value of by-product	1794 (6.00)	6252 (6.18)	10531 (5.38)	8522 (1.75)	15205 (1.18)	7581 (3.05)
Total value	29927 (100.00)	101243 (100.00)	195750 (100.00)	486393 (100.00)	1290691 (100.00)	248814 (100.00)
Rs/ha						
Value of product	63443 (94.00)	73428 (93.82)	72109 (94.62)	82740 (98.25)	82924 (98.82)	78584 (96.95)
Value of by-product	4047 (6.00)	4833 (6.18)	4100 (5.38)	1476 (1.75)	989 (1.18)	2470 (3.05)
Total value	**67490[b] (100.00)**	**78260[a] (100.00)**	**76209[a] (100.00)**	**84215[a] (100.00)**	**83912[a] (100.00)**	**81053 (100.00)**

Note: Figures in the parentheses are percentage to the total value.

[a] same alphabet signifies no significant difference among categories with same alphabet.

[b] different alphabet signifies significant difference from category with different alphabet.

It revealed that only small farmers showed positive and significant correlation between farm-size and productivity. For all other categories, there was neutral correlation between farm-size and productivity (non-significant coefficients) (Table 7.4). Regression equation model was applied using land under cultivation in kharif season (ha) as independent variable and productivity of the farm (Rs/ha) in kharif season as dependent variable. It was observed that both the intercept and the coefficient were significant and positive. Overall equation was significant but the R-square was very low (2.3%). It depicts that there was positive and significant relation between the farm-size and productivity but there were more important factors other than farm-size which affects the productivity. In nutshell, it was confirmed the relationship between farm-size and productivity for kharif season was non-negative i.e., either neutral or positive. It is also confirmed that productivity of kharif crops was lower on marginal farms than all other farms. Moreover, the positive relationship within small farm-size category was visualized as bridge between low productivity group (marginal farms) and high productivity groups (small, semi-medium, medium and large farms). Thereafter, the relationship remained neutral for high productivity group as correlation within these categories being non-significant and ANOVA also revealing no significant productivity difference between these categories.

Table 7.4: Regression and correlation coefficients for farm-size and productivity, kharif season, Punjab

Particulars	Marginal	Small	Semi-medium	Medium	Large	Overall
Correlation						
Coefficient	0.161	0.311	0.052	0.103	0.005	0.152
p-value	0.112	0.001	0.505	0.266	0.984	0.001
Regression (Overall)						
Coefficient (B_1)	925.54					
R^2	0.023					
p-value	0.001					

7.2.2 Farm-size and productivity relationship-Rabi crops

During rabi season, wheat was the main crop in Punjab. There were 556 observations for rabi season. Outliers were removed using interquartile range on value productivity. The final sample had 525 farmers sowing rabi crops, out of which 98 farmers belonged to marginal category, 130 farmers were small, 158 farmers were semi-medium, 118 farmers were medium and 21 were large farmers. The total value of crop on marginal farms was Rs 26254 and on large farms was Rs 10.13 lakh (Table 7.5). Overall, the value of the crop on all farm-size categories in rabi season was Rs 2.01 lakh. The overall value productivity of rabi crops for all the farm-size categories was Rs 67468. It was found that there was upward trend for farm-size and productivity relationship. The productivity of marginal farms was the lowest (Rs 59052) and the large farms was the highest (Rs 73675) for the rabi season. Overall correlation coefficient was worked between farm-size and productivity which was found to be positive (0.20) and significant (Table 7.6). By using ANOVA, it was found that the productivity on all the farm-size categories was identical except marginal farms where productivity was relatively lower. The correlation coefficient between farm-size and productivity was analyzed within each farm-size category. It was found that the correlation coefficient was non-significant within marginal, small and medium farm-size categories. Moreover, the correlation coefficient for semi-medium category was weak (r=0.14) and significant (p=0.08). The value for correlation coefficient was strongly positive (r=0.69) and significant for large farms. Regression equation model was applied using land under cultivation in rabi season (ha) as independent variable and productivity of the farm in rabi season (Rs/ha) as dependent variable. It was observed that the intercept and coefficient were significant. Overall regression equation was significant and the R-square was very low. It can be deduced that there is no significantly negative linear relationship between farm-size and productivity observed during rabi season.

Table 7.5: Gross returns from rabi crops across different farm-size categories, Punjab

(in Rs)

Particulars	Marginal	Small	Semi-medium	Medium	Large	Overall
per farm						
Value of product	24094 (91.77)	74037 (90.71)	154706 (89.48)	335014 (90.68)	941004 (92.93)	182328 (90.86)
Value of by-product	2160 (8.23)	7578 (9.29)	18179 (10.52)	34422 (9.32)	71542 (7.07)	18349 (9.14)
Total value	**26254 (100.00)**	**81615 (100.00)**	**172885 (100.00)**	**369436 (100.00)**	**1012546 (100.00)**	**200677 (100.00)**
per ha						
Value of product	54193 (91.77)	58265 (90.71)	60012 (89.48)	60181 (90.68)	68470 (92.93)	61299 (90.86)
Value of by-product	4859 (8.23)	5964 (9.29)	7052 (10.52)	6183 (9.32)	5206 (7.07)	6169 (9.14)
Total value	**59052[b] (100.00)**	**64229[a] (100.00)**	**67046[a] (100.00)**	**66364[a] (100.00)**	**73675[a] (100.00)**	**67468 (100.00)**

Note: Figures in the parentheses are percentage to the total value.
a same alphabet signifies no significant difference among categories with same alphabet.
b different alphabet signifies significant difference from the category with different alphabet.

Table 7.6: Regression and correlation coefficient for farm-size and productivity, rabi crops, Punjab

Particulars	Marginal	Small	Semi-medium	Medium	Large	Overall
Correlation						
Coefficient	0.100	0.095	0.139	0.065	0.686	0.195
P-value	0.328	0.281	0.081	0.486	0.001	0.001
Regression (Overall)						
Coefficient (B_1)	888.32					
R^2	0.038					
p-value	0.001					

7.2.3 Farm-size and productivity relationship-All crops

The annual value of production and productivity of crops cultivated by the farm households during survey year was under consideration in this section (Table 7.7). The sample size for the Punjab was 558 farm households. Per annum value of output per hectare was computed. The data was scrutinized for potential outliers.

Table 7.7: Gross returns across different farm-size categories, Punjab

(Rs/annum)

Particulars	Marginal	Small	Semi-medium	Medium	Large	Overall
Per farm						
Value of product	54852 (93.40)	166750 (92.44)	343793 (92.39)	818148 (95.04)	2216490 (96.23)	423998 (94.32)
Value of by-product	3873 (6.60)	13639 (7.56)	28320 (7.61)	42704 (4.96)	86746 (3.77)	25550 (5.68)
Total value	**58725 (100.00)**	**180389 (100.00)**	**372113 (100.00)**	**860852 (100.00)**	**2303237 (100.00)**	**449548 (100.00)**
per ha						
Value of product	113717 (93.40)	119486 (92.44)	124690 (92.39)	136409 (95.04)	149657 (96.23)	133258 (94.32)
Value of by-product	8029 (6.60)	9773 (7.56)	10271 (7.61)	7120 (4.96)	5857 (3.77)	74195 (5.68)
Total value	**121747[b] (100.00)**	**129259[ab] (100.00)**	**134962[a] (100.00)**	**143529[a] (100.00)**	**155514[a] (100.00)**	**141288 (100.00)**

Note: Figures in the parentheses are percentage to the total value.
a same alphabet signifies no significant difference among categories with same alphabet.
b different alphabet signifies significant difference from category with different alphabet.

The outliers were detected and removed. The final sample retained was 539 farm households with 102 (18.92%) marginal households, 133 (24.68%) small households, 163 (30.24%) semi-medium households, 120 (22.26%) medium households and 21 (3.90%) large households. The overall gross returns per annum for a farm in Punjab were Rs 4.50 lakh. The value of product per farm and the value of by-product per farm increased with increase in farm-size. The gross returns for marginal farms were Rs 58,725 and for large farms were Rs 23.03 lakh. The gross returns per hectares for different farm-size categories was also analyzed to examine the productivity of different farms. The overall productivity per year for all farm-size categories was Rs 1.41 lakh. The productivity of the large farms was the highest (Rs 1.56 lakh), followed by medium farms (Rs 1.44 lakh), semi-medium farms (Rs 1.35 lakh), small farms (Rs 1.29 lakh) and the lowest (Rs 1.22 lakh) for marginal farms. The value of correlation coefficient between farm-size and productivity was positive (r=0.21) and significant (Table 7.8).

Table 7.8: Regression and correlation coefficient for farm-size and productivity, all crops, Punjab

Particulars	Marginal	Small	Semi-medium	Medium	Large	Overall
Correlation						
Coefficient	0.167	0.062	0.028	0.084	0.526	0.207
P-value	0.093	0.478	0.727	0.362	0.014	0.001
Regression (Overall)						
Coefficient (B_1)	2529.06					
R^2	0.043					
p-value	0.001					

The results of ANOVA and post-hoc test reveals that there were broadly two productivity groups in Punjab agriculture. First group consists of marginal farmers for whom the productivity was comparatively lower while the other group was comprised of farmers belonging to semi-medium, medium and large farm-size categories with high productivity. The productivity on small farm-size category was somewhat similar to both the productivity groups. The correlation coefficient for within each farm-size category was positive and significant for both ends of farm-size classification i.e., marginal and large farm-size categories while non-significant for small, semi-medium and medium farm-size categories. Regression equation model was also applied using operational holdings as independent variable and productivity of the farm as dependent variable. It was observed that the intercept and coefficient were significant and positive. Overall equation was significant and the R-square was very low (4.28%). It can be confirmed that marginal farms have the lowest productivity among different farm-size categories. The productivity for small, semi-medium and medium farms was size neutral. The productivity of the largest farms within the large category was the highest (intra-farm positive correlation for large farms). Hence, the overall trend seems to be positive and it was confirmed that inverse relationship between farm-size and productivity was absent in Punjab. The relationship has been found opposite to the one reported for overall country. Marginal and small farmers were found to be more productive than larger farm size categories (Singh et al 2011). However, the farm-size and productivity relationship for Punjab agriculture was found to be positive in earlier studies also (Rudra and Chattopadhyay 1976, Bhalla and Chadha 1982, Singh et al 2000).

7.3 RELATIONSHIP BETWEEN INPUT-USE AND FARM-SIZE

Further, analysis was carried to examine the relationship between different input-use per hectare and farm-size. Various inputs included were cost of fertilizers & manures, plant protection chemicals, hired labour, diesel, repair & maintenance and cost of hiring machinery. The relationship between farm-size and input-use for different farm-size categories in Punjab has been presented in Table 7.9.

Table 7.9: Correlation and regression coefficients between farm-size and input-use, Punjab

Inputs	Correlation coefficient	Regression coefficient	R^2
Fertilizers & manures	0.159 (0.001)	209 (0.001)	0.025
Plant protection chemicals	0.194 (0.001)	236 (0.001)	0.038
Hired labour	0.255 (0.001)	446 (0.001)	0.065
Diesel	0.174 (0.001)	285 (0.001)	0.030
Repairs & maintenance	-0.02 (0.57)	-24 (0.57)	0.001
Cost of hiring machinery	-0.099 (0.022)	-114 (0.02)	0.010

The correlation coefficient for use of fertilizers & manure the relationship was positive (0.16), for plant protection chemicals was positive (0.19), for hired labour the relationship was positive (0.25), for diesel the relationship was positive (0.17), for repairs & maintenance was negative (-0.02) and for hiring machinery was negative (-0.099). The findings are supported by the earlier studies (Bhalla and Chadha 1982) in which it was found that fertilizers & manure and chemicals use per cropped area were positively related to farm-size.

All these inputs had significant relationship with farm-size except repairs & maintenance. Regression equation was fitted using operational holdings as independent variable and input-use per hectare as dependent variable. The coefficient for use of fertilizers & manure the relationship was positive (299), for plant protection chemicals was positive (236), for hired labour the relationship was positive (446), for diesel the relationship was positive (285), for repairs & maintenance was negative (-24) and for hiring machinery was negative (-114). All these inputs had significant relationship with farm-size except per hectare cost on repairs & maintenance. The study conducted by Rudra and Chattopadhyay 1976, Bhalla and Chadha 1982 found negative correlation between farm-size and per hectare total input-use.

The perusal of farm-size and productivity relationship confirmed that there is no inverse relationship between farm-size and productivity in kharif as well as rabi crops. Overall, the absence of inverse farm-size and productivity relationship was confirmed for Punjab. It was also found that there was no inverse relationship in paid-out cost and there is strong evidence of positive correlation between farm size and paid-out cost above two hectares of farm-size. Farm-size and input-use relationship was positive for fertilizer & manures, plant protection chemicals, hired labour, diesel except per hectare cost on hiring machinery and repairs & maintenance. The relationship was negative for hiring machinery and non-significant for repairs & maintenance.

Chapter-8

Level and Pattern of Income Among Farm Households in Punjab

The information on the sources-wise income across different farm-size categories have been presented in this chapter. The sample size for the Punjab in the present study was 558 farm households. The data was scrutinized for potential outliers. There were few households in which the livestock income appeared to be extremely odd. Those households were subjectively removed and the final sample size was 553 household with 108 (19.53%) marginal households, 142 (25.68%) small households, 165 (29.84%) semi-medium households, 118 (21.34%) medium households and 20 (3.62%) large households. It has been observed that the households engaged in agricultural activities often practice pluriactivity. They either engaged in agricultural allied enterprises or pursue non-farm activities. The sources of incomes enlisted in the present survey were crops, livestock, wages, salaries and non-farm business.

8.1 SOURCES OF INCOME AMONG FARM HOUSEHOLDS

The information on the engagement of households in different economic activities is given in the Table 8.1. As the basis of selection was crop production, all the households were engaged in this activity. Most of the farm households were also engaged in livestock production (94.03%). There were few farm households whose members worked on casual wages (2.89%) mainly in marginal farm-size category. Out of total households, 15.73 per cent and 8.32 per cent households were receiving income from regular wages/salaries and non-farm business. There were few large farm-size category households engaged in casual wages, regular wages/salaries and non-farm business. The per cent share of farm households engaged in non-agricultural activities i.e., wages, salaries and non-farm business was the highest for households having marginal farm-size of operational holdings. The second highest was for the households belonging to small farm-size category. It showed that the sources of income of marginal farmers were more diversified than other categories of farm households.

Table 8.1: Distribution of farm households according to sources of income, Punjab

(Number of households)

Sources of income	Farm-size categories					Overall
	Marginal	Small	Semi-medium	Medium	Large	
Income from crops	108 (100.00)	142 (100.00)	165 (100.00)	118 (100.00)	20 (100.00)	553 (100.00)
Income from livestock	102 (94.44)	129 (90.85)	152 (92.12)	117 (99.15)	20 (100.00)	520 (94.03)
Casual wages	12 (11.11)	2 (1.41)	1 (0.61)	1 (0.85)	-	16 (2.89)
Regular wages and salaries	26 (24.07)	25 (17.61)	17 (10.30)	17 (14.41)	2 (10.00)	87 (15.73)
Non-farm business	14 (12.96)	19 (13.38)	6 (3.64)	6 (5.08)	1 (5.00)	46 (8.32)
Number of households	**108 (100.00)**	**142 (100.00)**	**165 (100.00)**	**118 (100.00)**	**20 (100.00)**	**553 (100.00)**

Note: Household members may engage in more than one source of income

8.2 EXTENT OF PLURIACTIVITY AMONG FARM HOUSEHOLDS

Pluriactivity is defined as the participation of any member of farm household in income earning activity either on-farm or off-farm in order to secure the economy and welfare of the household. The analysis for status of pluriactivity among farm households was worked. The number of farm households were tabulated according to number of income sources of the household and presented in the Table 8.2. It was found that majority of the farm households were engaged in pluriactivity (95.84%). Only 4.16 per cent households were dependent on single source of income. Most of the farm households were earning household income from two sources (72.51%). There were 23.33 per cent farm households which had more than two income sources. In different farm-size categories, the farm household above two hectares of operational holdings were mainly dependent on two income sources with more than 80 per cent households having two income sources. These sources of income were mainly crop production and livestock i.e., income from agriculture. In households with less than two hectares of operational holdings, there was variation in the number of income sources of the households. In marginal category, 53.70 per cent farmers have two income sources, 38.89 per cent households have three income sources, a few households even have four income sources also (3.70%).

In small farm-size category, 7.04 per cent farm households have single income source, 64.79 per cent farm households have two income sources, 26.06 per cent households have three income sources, a few households even have four income sources also (2.11%).

Table 8.2: Distribution of farm households according to number of income sources, Punjab

(Number of households)

Number of income sources	Farm-size categories					
	Marginal	Small	Semi-medium	Medium	Large	Overall
One	4 (3.70)	10 (7.04)	8 (4.85)	1 (0.85)	-	23 (4.16)
Two	58 (53.70)	92 (64.79)	139 (84.24)	95 (80.51)	17 (85.00)	401 (72.51)
Three	42 (38.89)	37 (26.06)	17 (10.30)	20 (16.95)	3 (15.00)	119 (21.52)
Four	4 (3.70)	3 (2.11)	1 (0.61)	2 (1.69)	-	10 (1.81)
Total	108 (100.00)	142 (100.00)	165 (100.00)	118 (100.00)	20 (100.00)	553 (100.00)

Note: Figures in the parentheses are percentage to the total.

8.3 SOURCE-WISE LEVEL OF INCOME OF FARM HOUSEHOLDS

The average income earned by the farm households was 3.26 lakhs in which the major contributor was crops (Rs 2.67 lakh). It was followed by regular wages/ salaries (Rs 28263), livestock (Rs 21550), non-farm business (Rs 7571) and casual wages (Rs 1639). In different farm size categories, the income of the farm households increased with the increase in the farm-size. The average income per households was Rs 1.19 lakh, Rs 2.08 lakh, Rs 3.08 lakh, Rs 5.43 lakh and Rs 11.47 lakh for marginal, small, semi-medium, medium and large farm-size categories, respectively. The participation of farm households in casual wages was minimal except in marginal farm households. The income per households from casual wages was Rs 6100, Rs 1035, Rs 152 and Rs 640 for marginal, small, semi-medium and medium categories, respectively. Similarly, the participation of farm households in regular wages/salaries was relatively higher. The income per households from regular wages/salaries was Rs 31385, Rs 38127, Rs 22088, Rs 24154 and 16550 for marginal, small, semi-medium, medium and large farm-size categories, respectively. Source-wise income earned by the farm households across farm-size categories have been presented in Table 8.3.

Table 8.3: Level of income of the farm households across different farm-size categories, Punjab

(Rs/household/annum)

Income sources	Farm-size categories					Overall
	Marginal	Small	Semi-medium	Medium	Large	
a) Crops	47083	152758	259666	480053	1063950	266812
b) Livestock	24772	11860	20977	25940	51791	21550
c) Casual wages						
i) Agriculture	1343	380	152	640	-	542
ii) Construction	3619	655	-	-	-	875
iii) Others	1139	-	-	-	-	222
Total (i+ii+iii)	**6100**	**1035**	**152**	**640**	-	**1639**
d) Regular wages/salary						
i) Agriculture	519	296	194	3261	16550	1529
ii) Manufacturing	6579	2465	679	-	-	2120
iii) Services	24288	35366	21215	20893	-	24613
Total (i+ii+iii)	**31385**	**38127**	**22088**	**24154**	**16550**	**28263**
e) Non-farm business	9740	4515	4716	11996	15000	7571
Total (a+b+c+d+e)	**119081**	**208295**	**307598**	**542783**	**1147290**	**325835**

8.4 PER CAPITA INCOME OF FARM HOUSEHOLDS

The annual income earned by the farm households on per capita basis across different farm-size categories have been presented in the Table 8.4. The annual average income per capita earned by the farm households was Rs 59428 in which the major contributor was crops (Rs 48663). It was followed by regular wages/ salaries (Rs 4489), livestock (Rs 3930), non-farm business (Rs 1381) and casual wages (Rs 299). In different farm size categories, the per capita income of the farm households increased with the increase in the farm-size. The average income per households was Rs 24732, Rs 41252, Rs 56708, Rs 84720 and Rs 159346 for marginal, small, semi-medium, medium and large farm-size categories, respectively. In salary and wages categories, the farm households were earning decent income from services sector only.

The income per households from casual wages was Rs 1267, Rs 205, Rs 28 and Rs 100 for marginal, small, semi-medium and medium categories, respectively. Similarly, the participation of farm households in regular wages/salaries was relatively higher. The income per households from regular wages/salaries was Rs 6518, Rs 7551, Rs 4072, Rs 3770 and 2299 for marginal, small, semi-medium, medium and large farm-size categories, respectively.

Table 8.4: Source-wise per capita income of the farm household across different farm-size categories, Punjab

(Rs/capita/annum)

Income sources	Farm-size categories					Overall
	Marginal	Small	Semi-medium	Medium	Large	
a) Crops	9779	30253	47871	74929	147771	48663
b) Livestock	5145	2349	3867	4049	7193	3930
c) Casual wages						
i) Agriculture	279	75	28	100	-	99
ii) Construction	752	130	-	-	-	160
iii) Others	237	-	-	-	-	40
Total (i+ii+iii)	**1267**	**205**	**28**	**100**	-	**299**
d) Regular wages/salary						
i) Agriculture	108	59	36	509	2299	279
ii) Manufacturing	1366	488	125	-	-	387
iii) Services	5044	7004	3911	3261	-	4489
Total (i+ii+iii)	**6518**	**7551**	**4072**	**3770**	**2299**	**5155**
e) Non-farm business	2023	894	869	1872	2083	1381
Total (a+b+c+d+e)	**24732**	**41252**	**56708**	**84720**	**159346**	**59428**

8.5 SHARE OF DIFFERENT INCOME SOURCES IN TOTAL INCOME

The income earned by the farm households in different farm-size categories have been presented in the Table 8.5. The major contributor to total income was crops (81.89%). It was followed by regular wages/salaries (8.67%), livestock (6.61%), non-farm business (2.32%) and casual wages (0.50%). Across different farm-size categories, the share of income from crops increased with increase in farm-size. It was 39.54 per cent for marginal, 73.34 per cent for small, 84.42 per cent for semi-medium, 88.44 per cent for medium and 92.74 for large farm households. The per cent share of income from livestock in total income decreased with increase in farm-size categories.

The per cent share was 20.80 per cent for marginal, 5.69 per cent for small, 6.82 per cent for semi-medium, 4.78 per cent for medium and 4.51 per cent for large farm households. Same trend was observed in all other income sources i.e., casual wages, regular wages/salaries and non-farm business which showed inverse relationship with farm-size categories. For regular wages/salaries, the per cent share was 26.36 per cent for marginal farm households which decreased to 1.44 per cent for large farm households. Similarly, the percent share for non-farm business was 8.18 per cent for marginal farm households decreased continuously and was 1.31 per cent for large farm size households.

Table 8.5: Share of different income sources in total income across different farm-size categories, Punjab

(Per cent)

Income sources	Farm-size categories					Overall
	Marginal	Small	Semi-medium	Medium	Large	
a) Crops	39.54	73.34	84.42	88.44	92.74	81.89
b) Livestock	20.80	5.69	6.82	4.78	4.51	6.61
c) Casual wages						
i) Agriculture	1.13	0.18	0.05	0.12	-	0.17
ii) Construction	3.04	0.31	-	-	-	0.27
iii) Others	0.96	0.00	-	-	-	0.07
Total (i+ii+iii)	**5.12**	**0.50**	**0.05**	**0.12**	-	**0.50**
d) Regular wages/salary						
i) Agriculture	0.44	0.14	0.06	0.60	1.44	0.47
ii) Manufacturing	5.52	1.18	0.22	-	-	0.65
iii) Services	20.40	16.98	6.90	3.85	-	7.55
Total (i+ii+iii)	**26.36**	**18.30**	**7.18**	**4.45**	**1.44**	**8.67**
e) Non-farm business	8.18	2.17	1.53	2.21	1.31	2.32
Total (a+b+c+d+e)	**100.00**	**100.00**	**100.00**	**100.00**	**100.00**	**100.00**

ITo conclude, it was found that the main source of farm households was crops (81.89%), followed by regular wages/ salaries (8.67%), livestock (6.61%) and non-farm business (2.32%). The sources of income for marginal farmers are completely different from rest of the farm-size categories. The share crops income was 39.54 per cent, followed by regular wages/ salaries (26.36%), livestock (20.80%), non-farm business (8.18%) and casual wages (5.12%). Only the marginal farm-size worked as casual workers and earned income as casual wages.

Chapter-9

Income Distribution and Inequality Among Farm Households in Punjab

Inequality is defined as the extent of unequal distribution of resources among the members of the society. The analysis of the distribution of income provides the information on the extent of inequality among the farm households in Punjab. As discussed earlier, income inequality has negative effects on both macro and micro level. At macro level, it hampers growth and at micro level it affects the welfare of the households belonging to the lower income groups. The Chapter is decided into 2 sub-heading as under:

9.1 Distribution of income among farm household

9.2 Decomposition of income inequality among farm households

9.1 DISTRIBUTION OF INCOME AMONG FARM HOUSEHOLDS

Distribution of income among household is best depicted with the use of Lorenz curve. The Lorenz curve for farm household is reputed for overall income distribution as well as for individual income source and farm household is discussed as under:

9.1.1 Lorenz curve for overall income of the farm households

The Lorenz curve for total income received by all the selected farm households. has been presented in the Figure 2. It depicted the distribution of income among the farm households. The lower decile group has less than one per cent share in total income. Similarly, the bottom 20 per cent had only 2.65 per cent share in total income. It clearly shows highly skewed income distribution among farm households.

The lower 40 per cent of the farm households had access to 10.83 per cent of the total income among farm households.

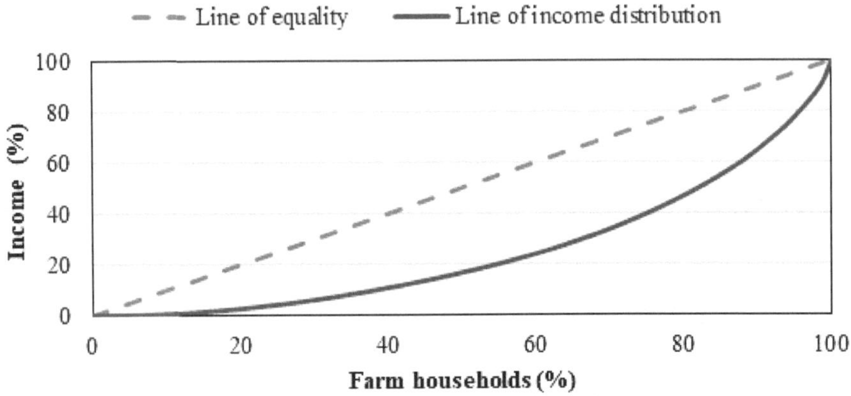

Figure 2: Lorenz curve: Distribution of farm households according to income of the farm household

The bottom 60 per cent farm households had around 24 per cent share in total income distribution. If we observed the upper part of the population, upper 20 per cent farm households had 52.98 per cent share in the total income distribution. the upper decile farm household population had access to around 35 per cent population and top one farm households earned around 7.5 per cent of the total income.

9.1.2 Lorenz curve for each income source

Source-wise income distribution has been presented in Figure 3. Different sources of income of the households includes income from crop cultivation, income from livestock, income from casual wages, income from regular wages/salary and non-farm business income. The income from each source was arranged in ascending order and then Lorenz curve for each source was derived. It was found that for income from crop production has almost same distribution as the total income. It was mainly due the highest share of crop income in the total income as found in Section 4.4. The bottom 20 per cent farm households had only 1.4 per cent share in crop income. Bottom 40 per cent had 8.18 per cent share in the crop income. It can be observed that the crop income was highly unequally distributed among farm households. The top 20 per cent farm households had access to 56.75 per cent of the total income. Top decile population has access to around 39 per cent share of the total income. The income from the livestock was highly unequally distributed. Top one per cent of the farm households have around 65 per cent share in livestock income. The top five per cent population have access to 80 per cent share of the livestock income. The distribution of bottom 90 per cent farm households had only 13.33 per cent income share in livestock income. The income from casual wages also followed similar pattern but we know that there were few households engaged in casual labour so the skewed distribution was expected. Only 3.72 per cent farm households earned income through in casual wages. Income from regular wages and salaries was available 14.57 per cent population only, hence, the distribution was skewed. The income distribution of non-farm business was also skewed. Less than 10 per cent households have access to non-farm business. Top 5 per cent population had access to 90 per cent share in non-farm business income.

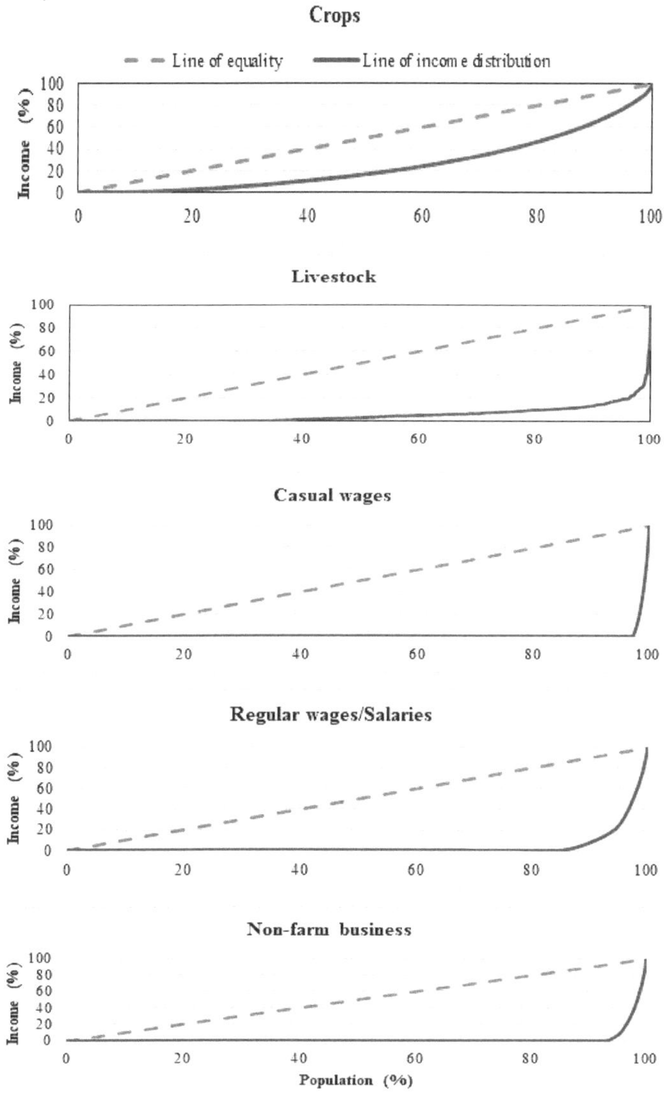

Figure 3: Lorenz curves for different income sources

9.1.3 Distribution of farm households in different income quintiles

Income quintiles were formed on the basis of the total income of the household. The Table 9.1 provides the information on the number of households in each quintile in different farm size groups. The large farmers were concentrated in the fourth and the highest income quintiles while the marginal farmers were concentrated in the lower quintiles with around 50 per cent of the marginal farm households in bottom quintile and 33.33 per cent is second quintile. The distribution of small farm households in different income quintiles was somewhat similar to marginal households while the distribution of medium farm households in different income quintiles was somewhat similar to large households.

Table 9.1: Farm-size category-wise number of farm households in each quintile

(Number)

Quintile	Farm-size categories				
	Marginal	**Small**	**Semi-medium**	**Medium**	**Large**
Bottom	52 (48.15)	36 (25.35)	19 (11.52)	3 (2.54)	-
Q2	36 (33.33)	47 (33.10)	21 (12.73)	7 (5.93)	-
Q3	11 (10.19)	35 (24.65)	52 (31.52)	12 (10.17)	-
Q4	6 (5.56)	13 (9.15)	55 (33.33)	34 (28.81)	4 (20.00)
Top	3 (2.78)	11 (7.75)	18 (10.91)	62 (52.54)	16 (80.00)
Overall	**108 (100.00)**	**142 (100.00)**	**165 (100.00)**	**118 (100.00)**	**20 (100.00)**

Note: Figures in the parentheses are percentage to the overall.

9.1.4 Distribution of income of farm households according to quintile groups

The income of the farm households was divided into five equal quintiles formed on the basis of total income of the farm household. The information has been presented in Table 9.2.

In each successive quintile, the income got double as two subsequent quintiles were compared, except bottom quintile where the difference was four times i.e., the average income for first quintile was Rs 33075 while that for second quintile as Rs 1.33 lakh. For quintile three, four and five the average income was Rs 2.18 lakhs, Rs 3.75 lakhs and Rs 8.74 lakhs, respectively. The per cent share of income was worked for individual income source.

The income share of different sources in total income for bottom quintile was the highest for crops (104.4%) followed by regular salaries (17.6%), casual wages (9.19%), non-farm business (7.72%). The livestock income was negative in bottom quintile. The next three quintiles were having the highest share of crop income. It was followed by income from livestock. Similarly, the regular wages category has the next highest share in each quintile following the livestock income. It was followed by non-farm business and casual wages. The top quintile was having the highest crop (81.00%). The second highest source of income was the regular wages/salaries (10.93%), followed by livestock (5.63%) and non-farm business (2.38%) and casual wages (0.06%). The income was highly unequally distributed among the farm households. Various sources of income show varied share in each income quintile. Thus, further decomposition of can be done to enquire in details of income distribution.

Table 9.2: Distribution of income of farm households according to quintile groups

Quintiles	Income sources					Total income
	Crops	Livestock	Casual wages	Regular wages/ Salaries	Non-farm business	
Q1	34521 (104.37)	-12861 (-38.88)	3041 (9.19)	5823 (17.6)	2553 (7.72)	33075 (100.00)
Q2	99100 (74.48)	15678 (11.78)	3935 (2.96)	9931 (7.46)	4406 (3.31)	133049 (100.00)
Q3	179899 (82.61)	18589 (8.54)	709 (0.33)	12115 (5.56)	6459 (2.97)	217772 (100.00)
Q4	316062 (84.15)	37398 (9.96)	-	18408 (4.90)	3735 (0.99)	375603 (100.00)
Q5	707665 (81.00)	49171 (5.63)	491 (0.06)	95498 (10.93)	20810 (2.38)	873634 (100.00)
Overall	**266812 (81.89)**	**21550 (6.61)**	**1639 (0.50)**	**28263 (8.67)**	**7571 (2.32)**	**25835 (100.00)**

Note: Figures in the parentheses are percentage to the total income.

The digest of this section concludes huge income gap between upper and lower strata income. The distribution of income was highly skewed. It was found that the households in lower strata were mostly marginal and small while the large farm-size categories were mainly concentrated on upper strata. The income sources of the lower quintiles were diversified while that of the upper quintiles was concentrated on agriculture.

9.2 DECOMPOSITION OF INCOME INEQUALITY AMONG FARM HOUSEHOLDS

Most of the studies in India pertaining to income inequality of rural households and distributional consequences of income sources have been carried out at India level. However, India is a diverse country with large chunk of its population living in rural areas. There also exists high agro-climatic as well as cultural diversity in India. Regional studies in inequality will be useful to understand regional composition of income earned by agricultural households which makes large portion of rural population which may help to address the prevailing problems in particular region. The present section has been devoted to gain insights about income inequality among farm households in Punjab.

9.2.1 Income inequality among farm households

Income inequality among farm households has been analyzed by estimating Gini coefficient. Further, decomposition of Gini coefficient was done presenting source-wise inequality. The vertical decomposition of inequality among different income sources was done using Lerman and Yitzhaki (1985). Following Bellu and Liberati (2006), negative values were given small positive value to work out decomposition of income inequality. In present study the value used was 10-5. The major findings are presented in the Table 9.3. The Gini coefficient for income inequality among the farm households was found to be 0.50 which depicted presence of high income inequality among farm households. The analysis revealed that the income from crop production has the major share in total income while income from casual labour has the minimum share. The Gini coefficient for each individual source was high. It was highest for casual wages. It may be due to absence of people engaged in casual labour among farm households. Same will be the case for regular wages/salaries and non-farm business. Value of Gini correlation was high for crops (0.91). The Gini correlation for regular wages/salaries was also high (0.61). All other income sources have moderate correlation with the total income distribution. The correlation for casual wages was moderate but negative i.e., where total income increased the income from casual wages decreased and vice versa. The share of different sources of income in Gini coefficient was also worked out. It was found that crop production contributed the most (0.82). It was followed by regular wages/salaries (0.10), livestock (0.06), non-farm business activities (0.02). The casual wages as a source of income contributed negatively to the overall Gini coefficient though the contribution was almost negligible (0.004). Elasticities of the income sources were worked-out. The negative coefficient signifies that the income source is income inequality decreasing while the positive coefficient signifies that the source is inequality increasing. It was found that only two coefficients were significant i.e., livestock and casual wages. The coefficient of both the income sources were negative. It depicts that livestock and casual wages are income inequality decreasing sources of income.

These findings on inequality decreasing effect of livestock have been well established in literature (Anonymous, 2015a; Adams and He, 1995; Birthal et al 2002; Choudhary and Singh, 2020). All other coefficient were non-significant; hence, no conclusion can be drawn for them. In present situation, the availability of casual work can be enhanced to increase viability of household in lower strata. In long-run, an enquiry into the problems faced by the farm households in lower strata in fetching high salaried regular jobs can be explored and solution should be provided.

Table 9.3: Income inequality among farm households and its source-wise decomposition, Punjab

Income sources	**Share of source in total income (S_k)**	**Source Gini (G_k)**	**Gini correlation (R_k)**	**Share in Gini**	**Elasticity ()**
Crops	0.8178	0.5454	0.9142	0.8177	-0.0002 (0.0139)
Livestock	0.1008	0.6459	0.4358	0.0569	-0.0439* (0.0049)
Casual wages	0.0050	0.9815	-0.4369	-0.0043	-0.0093* (0.0029)
Regular wages & salary	0.0862	0.9290	0.6114	0.0981	0.0120 (0.0103)
Non-farm business	0.0273	0.9609	0.4368	0.0230	-0.0043 (0.0050)
Total income		**0.4987**			

Bootstrapped standard errors with 1000 replications in parentheses, *significant results

9.2.2 Horizontal decomposition of inequality by farm-size categories

Any inequality measure used must qualify the standard characteristics of inequality measures i.e., transferability, independence, additivity, sub-group consistency. Property of subgroup consistency: the property that if inequality increases in a population subgroup, then, other things being equal, inequality increases overall. It is possible to find cases where the Gini coefficient in a subgroup increases and the Gini coefficient overall falls, violating subgroup consistency. However, the Theil indices based on the entropy concept are naturally decomposable by population subgroup. This property does not depend on the additivity of independent events in the information function but because of the aggregation of entropy from the individual information components which induces additive separability (Cowell, 2005). The total income inequality as well as inequality in various sources of income were` decomposed into within and between sources categorized into different farm-size categories (Table 9.4). It was found that total income inequality was equally caused by within as well as between farm-size category inequality affect. In case of crops cultivation, the total income inequality was equally contributed by within and between groups equally. The coefficient of Theil index for crop production for within and between farm-size categories was found to be 0.266 and 0.278, respectively. On the contrary, high income inequality in livestock was more due to within categories inequality. There was comparatively lesser contribution of between categories inequality. The coefficient of Theil index for livestock production for within and between farm-size categories was found to be 0.772 and 0.034, respectively. Same pattern was observed for regular wages/salaries and non-farm business. The coefficient of Theil index for regular wages/salaries for within and between farm-size categories was 2.338 and 0.029, respectively. The coefficient of Theil index for non-farm business for within and between farm-size categories was 2.911 and 0.061, respectively. High income inequality in casual labour was high due to both between and within group income inequality. The coefficient of Theil index of income from casual wages for within and between farm-size categories was 3.023 and 0.737, respectively. In total, it was found that the income inequality was more due to within each category than the between farm size categories. The coefficient of Theil index for casual labour for within and between farm-size categories was found to be 0.266 and 0.179, respectively. The findings of the Theil index led to further enquire into the farm-size categories as the coefficient for within groups income inequality factors was quite high. The Gini coefficient for income inequality among the marginal, small, semi-medium, medium and large farm households was worked-out.

Table 9.4: Theil Index of inequality to analyze inequality between and within by farm-size categories

Income sources	**Farm-size categories**	
	Within	**Between**
Crops	0.2659	0.2765
Livestock	0.7721	0.0343
Casual wages	3.0231	0.7368
Regular wages & salary	2.338	0.0289
Non-farm business	2.9111	0.0605
Total income	**0.2655**	**0.1786**

9.2.3 Decomposition of income inequality within marginal farm households

The insights into decomposition of income inequality among the marginal farm households have been given in the Table 9.5.

The Gini coefficient was found to be 0.50 which depicts high income inequality among marginal farm households. Further, the analysis revealed that the income from crop production has the highest share in total income (0.39). it was followed by livestock (0.26), regular wages/salaries (0.26), non-farm business income (0.08). The share of income from casual wages was the lowest (0.05). The Gini coefficient for each individual source was high. It was the highest for casual wages (0.93) followed by non-farm business (0.93), regular wages/salaries (0.88), livestock (0.72) and crops (0.57). Value of Gini correlation was the highest for regular wages/salaries (0.79). It was followed by livestock (0.69), crops (0.60), non-farm business (0.52) and casual wages (0.21). The share of different sources of income in Gini coefficient was also worked out. The results were in parlance with the results of Gini correlation. The share of regular wages/salaries in total income inequality was the highest (0.36) and casual wages was the lowest (0.02). Elasticities of the income sources revealed that two sources of income were contributing significant to increase or decrease in inequality i.e., crops and regular wages/salaries for marginal farm households. The coefficient of crop income was negative while that of regular wages/salaries was positive.

Table 9.5: Decomposition of income inequality within marginal farm households in Punjab

Income sources	Share of source in total income (S_k)	Source Gini (G_k)	Gini correlation (R_k)	Share in Gini	Elasticity ()
Crops	0.3893	0.5702	0.6037	0.2684	-0.1209* (0.0529)
Livestock	0.2618	0.7203	0.6858	0.2590	-0.0028 (0.0416)
Casual wages	0.0502	0.9326	0.2080	0.0195	-0.0307 (0.1330)
Regular wages & salary	0.2583	0.8853	0.7932	0.3633	0.1050* (0.0525)
Non-farm business	0.0813	0.9274	0.5158	0.0779	-0.0034 (0.0202)
Total income		**0.4993**			

Bootstrapped standard errors with 1000 replications in parentheses, *significant results

It depicted that income from crops was inequality decreasing factor among the marginal farm households while regular wages and salaries was the income inequality increasing sources of income among marginal farm households. All other coefficient were non-significant; hence, no conclusion can be drawn for them. It is evident that the marginal households were the one lying in the lower income strata. It will be desirable to have more households in the upper tail of the distribution of income for marginal category. There can be further enquiry into the reasons leading few households to earn more from regular salaries while other depriving from deriving such benefits.

9.2.4 Decomposition of income inequality within small farm households

The decomposition of income inequality among the small farm households has been presented in Table 9.6. The Gini coefficient was found to be 0.46 which depicts high income inequality among small farm households. The share of different sources of income in total income was the highest for crop (0.73). It was followed by regular wages/salaries (0.18), livestock (0.10), non-farm business (0.05) and casual wages (0.005). The Gini coefficient for individual source was the highest for casual wages (0.99). It is due to engagement of less number of small farmers in this income source. Same pattern was observed for non-farm business (0.94) and regular wages/salaries (0.93). The Gini coefficient for income distribution livestock and crops among small farm households was 0.66 and 0.46, respectively. Value of Gini correlation was the highest for crops (0.83) and regular wages/salaries (0.83). It was followed by non-farm business (0.59), livestock (0.40) and casual wages (0.13). The results for share of different sources of income in overall Gini revealed that although the share of crop income in total income was high (0.73) but it contributed less to overall Gini (0.60). Similar was the case for livestock where share of crop income in total income was high (0.10) but it contributed less to overall Gini (0.06). On the contrary, the coefficient of share of regular wages/salaries in total income was 0.18 but it contributed more to overall Gini (0.30). Elasticities of income inequalities for different income sources revealed that three sources of income were contributing significant to increase or decrease in inequality i.e., crops livestock and regular wages/salaries. The coefficient of crop and livestock income was negative while that of regular wages/salaries was positive. It depicts that crop and livestock were inequality decreasing factor among the small farm households while regular wages and salaries was the income inequality increasing sources of income among small farm households. All other coefficient were non-significant; hence, no conclusion can be drawn for them. As the small households have characteristics resembling marginal households, it will be desirable to have more households in the upper tail of the distribution of income in this category also. Reasons causing few households to earn more from regular salaries while other depriving from deriving such benefits can be enquired further so that income base can be moved-up.

Table 9.6: Decomposition of income inequality within small farm households in Punjab

Income sources	Share of source in total income (S_k)	Source Gini (G_k)	Gini correlation (R_k)	Share in Gini	Elasticity ()
Crops	0.7301	0.4559	0.8282	0.5975	-0.1326* (0.0534)
Livestock	0.1003	0.6569	0.3961	0.0566	-0.0437* (0.0121)
Casual wages	0.0049	0.9878	0.1349	0.0014	-0.0035 (0.0058)
Regular wages & salary	0.1819	0.9263	0.8277	0.3023	0.1204* (0.0432)
Non-farm business	0.0467	0.9377	0.5868	0.0558	0.0090 (0.0119)
Total income		**0.4614**			

Bootstrapped standard errors with 1000 replications in parentheses, *significant results

9.2.5 Decomposition of income inequality within semi-medium farm households

The number of semi-medium and medium operational holdings are on the rise in Punjab. It would be interesting to see the composition of the income distribution among these categories. The Table 9.7 provides details about the decomposition of income inequality among the semi-medium farm households. The Gini coefficient was found to be 0.37 which depicts moderate income inequality among semi-medium farm households. The share of different sources of income in total income was the highest for crop (0.84). It was followed by livestock (0.11), regular wages/salaries (0.07), non-farm business (0.02) and casual wages (0.001). The Gini coefficient for individual source was the highest for casual wages (0.99). It is due to engagement of smaller number of semi-medium farmers in this income source. Same pattern was observed for non-farm business (0.98) and regular wages/salaries (0.94). The Gini coefficient for income distribution livestock and crops among semi-medium farm households was 0.61 and 0.35, respectively. The correlation of distribution of income sources with distribution of total income was also analyzed. Value of Gini correlation was the highest for crops (0.87) followed by regular wages/salaries (0.78), non-farm business (0.74) and livestock (0.45). The value of Gini correlation coefficient was highly negative for casual wages (-0.88).

Table 9.7: Decomposition of income inequality within semi-medium farm households in Punjab

Income sources	Share of source in total income (S_k)	Source Gini (G_k)	Gini correlation (R_k)	Share in Gini	Elasticity ()
Crops	0.8423	0.3549	0.8746	0.6979	-0.1445* (0.0522)
Livestock	0.1134	0.6117	0.4542	0.0841	-0.0293(0.0174)
Casual wages	0.0005	0.9939	-0.8780	-0.0011	-0.0016 (0.0013)
Regular wages & salary	0.0711	0.9436	0.7881	0.1410	0.0700* (0.0315)
Non-farm business	0.0152	0.9783	0.7409	0.0294	0.0142 (0.0110)
Total income		**0.3747**			

Bootstrapped standard errors with 1000 replications in parentheses, *significant results

The results for share of different sources of income in overall Gini revealed that although the share of crop income in total income was high (0.84) but it contributed relatively less to overall Gini (0.70). On the contrary, the coefficient of share of regular wages/salaries in total income was 0.07 but it contributed more to overall Gini (0.14). it was found that the income from crop was inequality decreasing factor for semi-medium farmers while the wages and salaries was inequality increasing factor.

9.2.6 Decomposition of income inequality within medium farm households

The vertical decomposition of income inequality among the medium farm households has been presented in Table 9.8. It was found that inequality among the income distribution among medium farms was very low. The value of Gini coefficient was found to be 0.30. It was found that the income of medium farms was highly concentrated on crop production (88.57%). It was followed by livestock (7.68%), regular wages/salaries (4.45%), non-farm business (2.21%) and casual wages (0.12%). The Gini coefficient for individual sources was the highest for casual wages (0.99). Same pattern was observed for non-farm business (0.97) and regular wages/salaries (0.93). The Gini coefficient for income from livestock and crops among medium farm households was 0.60 and 0.29, respectively. The correlation of distribution of income sources with distribution of total income was also analyzed. The value of Gini correlation was the highest for regular wages/salaries (0.93), followed by crops (0.90), non-farm business (0.84), livestock (0.29). The value of Gini correlation coefficient was negative and high for casual wages (-0.97). it was due to negligible presence of casual wage earners in larger farm-size categories. The results for share of different sources of income in overall Gini revealed that although the share of crop income in total income was high (0.89) but it contributed relatively less to overall Gini (0.76). Similarly, the share of livestock income in total income was 7.68 per cent but it contributed lesser to total income inequality (4.46%). On the contrary, the share of regular wages/salaries in total income was 4.45 per cent but it contributed more to overall Gini (9.38%). Consequently, it was found that the income from crop and livestock were inequality decreasing factor among medium farms while the regular wages/salaries was inequality increasing factor.

Table 9.8: Decomposition of income inequality within medium farm households in Punjab

Income sources	Share of source in total income (S_k)	Source Gini (G_k)	Gini correlation (R_k)	Share in Gini	Elasticity ()
Crops	0.8857	0.2876	0.9057	0.7606	-0.1251* (0.0409)
Livestock	0.0768	0.6010	0.2933	0.0446	-0.0322* (0.0140)
Casual wages	0.0012	0.9915	-0.9658	-0.0037	-0.0049 (0.0040)
Regular wages & salary	0.0445	0.9348	0.6841	0.0938	0.0493* (0.0026)
Non-farm business	0.0221	0.9709	0.8426	0.0596	0.0375 (0.0206)
Total income		**0.3033**			

Bootstrapped standard errors with 1000 replications in parentheses, *significant results

9.2.7 Decomposition of income inequality within large farm households

The vertical decomposition of income inequality among the large farm households has been presented in Table 4.6.7. It was found that inequality among the income distribution among large farms was contrary to medium farms. The value of Gini coefficient was found to be 0.41 depicting moderately high disparity in income among large farms. It was found that the income of large farms was highly concentrated on crop production (93.12%). It was followed by livestock (4.83%), regular wages/salaries (1.43%) and non-farm business (1.29%). The coefficient of individual sources of income was 0.95. 0.92, 0.41 and 0.44 for non-farm business, regular wages/salaries, livestock and crop income, respectively. The Gini correlation for different sources was the highest for crop income (0.99). Same pattern was observed for non-farm business (0.89) and regular wages/salaries (0.90). The results for share of different sources of income in overall Gini revealed that contribution of crop income to overall Gini was 92.06 per cent. Share of all other sources was less than two per cent. It was found that the livestock income was inequality decreasing factor for large farms.

Table 9.9: Decomposition of income inequality within large farm households in Punjab

Income sources	Share of source in total income (S_k)	Source Gini (G_k)	Gini correlation (R_k)	Share in Gini	Elasticity ()
Crops	0.9312	0.4088	0.9964	0.9206	-0.0106 (0.0153)
Livestock	0.0483	0.4429	0.3550	0.0184	-0.0299* (0.0148)
Casual wages	-	-	-	-	-
Regular wages & salary	0.0143	0.9258	0.8956	0.0287	0.0145 (0.0081)
Non-farm business	0.0129	0.9500	0.8947	0.0267	0.0138 (0.0121)
Total income		**0.4120**			

Bootstrapped standard errors with 1000 replications in parentheses, *significant results

The perusal of this section reveals that the inequality within farm size categories was higher than the inequality between categories. Further, enquiry into each category revealed that the inequality in all the categories have different sources. The extent of inequality was quite different for within all farm-size categories. The inequality decreased with increase in farm-size except large farm-size holdings. Regular wages came out to be inequality increasing factor in all categories except large category. Livestock came out to be inequality decreasing factor in overall decomposition as well as intra class decomposition of income inequality. This re-emphasis the importance of livestock sector in the agricultural economy of Punjab as found in India (Pavithra and Vatta 2013, Birthal et al 2014, Choudhary and Singh 2019) and the world (Lanjouw and Shariff 2004, Adams and He). Hence, livestock sector should be strengthened. There is need to explore the possibility to enhance the income from non-farm sources, especially non-farm business as the non-farm income sources have been identified as inequality decreasing sources (Pavithra and Vatta 2013, Adams and He 1995, Birthal et al 2014, Lanjouw and Shariff 2004).

Chapter-10

Summary, Conclusion, Suggestions and Policy Implications

Income inequality estimates are an important indicator to determine extent of overall inequality in the society. A welfare state tends to distribute income equally among the members of the society. It has been found that households in lower income group find difficult to stay healthy, accumulate physical and human capital which in turn affects growth (Galor and Moav 2004). Positive income distribution toward lower end is associated with higher Gross Domestic Product (GDP) growth and its absence destroys confidence of people in Government, increases corruption and nepotism. Knowledge of the income and various sources of the income and extent of inequality among the different farm size cultivators would be beneficial to work for welfare of the farmers in the state. Most of the studies on income distribution are based on secondary data (Birthal et al 2014, Chakravorty et al 2016). Also, income distribution and inequality in rural Punjab has been measured long ago (Bhalla and Chadha 1982, Singh 1985, Kaur 1992, Vatta 2006) and economy of Punjab has been through lot of transformation in this period. Studies on the level of income of rural households have been conducted but specific study on the income and its distribution among the farm households in Punjab are few. Thus, the specific objectives of the study were:

1. To analyze farm size and productivity relationship,
2. To estimate farm and non-farm income of different farm size categories,
3. To study income distribution among farmers; and
4. To analyze determinants of income inequality among farmers.

The data for the study were obtained from National Sample Survey Organization (NSSO) 70th round 'Situation Assessment Survey of Agricultural Households'. The data pertained to the agricultural year 2012-13. The survey covered 558 farm households. The relationship between farm-size and productivity was analyzed using various statistical techniques like correlation, regression and ANOVA. The income distribution was studied using quintiles groups and Lorenz curve. Gini coefficient and Theil index were worked to obtain coefficients of income inequality among farm households. Furthermore, income decomposition was done vertically as well as horizontally, using Theil index and Gini decomposition. The major findings of the study revealed that the average family size of the farm household was 5.49 members which increased with the farm size. The average operational holdings across different farm-size categories were 0.47, 1.40, 2.76, 6.00 and 14.81 hectares for marginal, small, semi-medium, medium and large farms, respectively. The share of leased-in land in the operational holding was about 25.72 per cent increased with increase in farm size. Most of the farm households belonged to general category (84.41%), followed by OBC (9.32), SC/ST group(6.27%). The farm households with the education level of graduation or above were more prevalent in semi-medium, medium and large farm-size categories. It was observed that majority of the farm household heads were aged between 30 to 60 years and around six per cent of the farm households were headed by female members. The main crops cultivated in Punjab were wheat and paddy occupying around 76 per cent area of cropped area. Other important crops were cotton (4.93%), maize (2.35%), potato (2.19%), sugarcane (1.25%). It was found that large farmers grow relatively more cash crops like sugarcane, potato, cotton than other farm-size categories while marginal category relied more on fodder crops. The cropping intensity had positive relationship with farm-size categories. The cropping intensity was 179.46, 179.71, 185.75, 188.73, 196.65 per cent for marginal, small, semi-medium, medium and large farm-size categories, respectively. The annual gross returns per farm for farm household were Rs 4.49 lakh. Across different farm-size categories, the gross returns were Rs 0.59 lakh, Rs 1.80 lakh, Rs 3.72 lakh, Rs 8.61 lakh and Rs 23.03 lakh for marginal, small, semi-medium, medium and large farm-size, respectively. Similarly, gross returns per hectare were analyzed and it was found to be the highest (Rs 1.56 lakh) for large farms, followed by medium farms (Rs 1.44 lakh), semi-medium farms (Rs 1.35 lakh), small farms (Rs 1.29 lakh) and the lowest (Rs 1.22 lakh) for marginal farms. The annual paid-out cost on crop cultivation includes cost of seeds, fertilizers, manures, plant protection chemicals, human labour, electricity, irrigation, diesel, repairs & maintenance, cost of hiring machinery, rent for leased-in land, interest on loans for cultivation purposes and other expenses. Structure of the paid-out cost revealed that the total cost per farm was Rs 1.98 lakh out of which the highest share was Rs 52061 (26.32%) for rent of leased-in land followed by cost of fertilizer & manures which was Rs 30192 (15.27%), human labour cost was Rs 27836 (14.07%). The other important expenses were diesel (11.43%), plant protection chemicals (10.40%) and cost of hiring machinery (7.73%), seed (5.42%) and minor repairs (3.20%). Other minor expenses and interest on loans for cultivation comprises of 6.16 per cent of the total paid-out cost. The details on paid-out costs per hectare across different farm-size categories revealed that the cost trends were upward for fertilizers & manure, human labour, diesel and plant protection chemicals. The per hectare cost of hiring machinery and seeds followed opposite trend. The overall value for gross returns, paid-out costs and returns over paid-out cost was Rs 4.49 lakh, Rs 1.98 lakh and Rs 2.52 lakh, respectively. The value of returns over paid-out cost was Rs 38116, Rs 1.12 lakh, Rs 2.26 lakh, Rs 4.83 lakh and 10.49 lakh for marginal, small, semi-medium, medium and large farm-size categories, respectively. Per capita returns over paid-out cost were analyzed and the overall value for per capita gross returns, paid-out costs and returns over paid-out cost was Rs 81585, Rs 35893 and 45691, respectively. Across different farm-size categories, the value of per capita returns over paid-out cost was Rs 7870, Rs 22347, Rs 41,322, Rs 75,672 and Rs 1,48,904 for marginal, small, semi-medium, medium and large farm-size categories, respectively. In per hectare terms, returns over paid-out cost had inverted U-shape. The large farmers earned the highest gross returns per hectare (Rs 155514) but the returns over paid-out cost were the lowest (Rs 70856) mainly due to higher proportion of leased-in rent cost. Farm-size and productivity relationship was analyzed at two levels i.e., productivity in physical terms and at farm level in value terms. Tools used to analyse farm-size and productivity were correlation coefficient and ANOVA. Regression analysis was also conducted to measure the magnitude of the farm-size and productivity relationship and to determine the goodness of fit. Crop-wise analysis for farm-size and productivity was conducted on wheat crop. The productivity of wheat per hectare was found to be 4.1, 4.3, 4.3, 4.5 and 4.6 tonnes for marginal, small, semi-medium, medium and large farm-size category, respectively. There was positive and significant correlation between area under wheat and productivity. Further, using ANOVA it was found that although the correlation was positive and significant, there was no significant difference in wheat productivity across different farm-size categories. Therefore, it was concluded that the difference in productivity was not high enough to categorize farm-size into different productivity groups. Intra-category correlation between farm-size and productivity revealed that the correlation coefficients for small and semi-medium were positive and significant while non-significant for all other categories. Regression equation was also fitted using area under wheat cultivation (ha) as independent variable and productivity of wheat crop (Kg/ha) as dependent variable. It was observed that the both intercept and coefficient were significant and positive. It was found that one hectare increase in crop may lead to 54 kg increase in productivity. Overall equation was significant but the R-square was very low which signifies that productivity was a significant factor explaining wheat productivity but there exist more important factors other than farm-size which determines wheat productivity. For kharif crops, the relationship between productivity and cropped area was found to be significant and positive. ANOVA analysis revealed that the productivity on marginal farms was different for all other farms. Hence, the productivity of kharif crops can be broadly categorized into two groups; one was marginal farms where productivity was lower and other group constituted small, semi-medium, medium and large farm-size categories where productivity was higher. The intra-category correlation for each farm-size category was also analyzed and results revealed that only small farmers showed positive and significant correlation between farm-size and productivity. For all other categories, there was neutral correlation between farm-size and productivity. Regression equation model was applied using land under cultivation in kharif season (ha) as independent variable and productivity of the farm (Rs/ha) in kharif season as dependent variable. It was observed that both the intercept and the coefficient were significant and positive. Overall equation was significant but the R-square was very low (2.3%). It depicts that there was positive and significant relation between the farm-size and productivity but there were more important factors other than farm-size which affects the productivity. Upward trend between farm-size and productivity relationship was also observed for rabi crops. Overall correlation coefficient was worked between farm-size and productivity which was found to be positive (0.20) and significant. ANOVA among different farm-size categories revealed that the productivity on marginal farms was relatively lower than other farm-size categories. It was found that the correlation coefficient was non-significant within marginal, small and medium farm-size categories. Moreover, the correlation coefficient for semi-medium category was weak (r=0.14) and significant (p=0.08). The value for correlation coefficient was strongly positive (r=0.69) and significant for large farms. Regression equation model was applied using land under cultivation in rabi season (ha) as independent variable and productivity of the farm in rabi season (Rs/ha) as dependent variable. It was observed that the intercept and coefficient were significant. Overall regression equation was significant and the R-square was very low. The value of production and productivity of all crops cultivated by the farm households during survey year were also analyzed. The overall productivity per year for all farm-size categories was Rs 1.41 lakh. The productivity of the large farms was the highest (Rs 1.56 lakh), followed by medium farms (Rs 1.44 lakh), semi-medium farms (Rs 1.35 lakh), small farms (Rs 1.30 lakh) and the lowest (Rs 1.22 lakh) for marginal farms. The value of correlation coefficient between farm-size and productivity was positive (r=0.21) and significant. it was found using ANOVA that there were broadly two productivity groups in Punjab agriculture. First group consists of marginal farmers for whom the productivity was comparatively lower while the other group was comprised of farmers belonging to semi-medium, medium and large farm-size categories with high productivity. The productivity on small farm-size category was somewhat similar to both the productivity groups. The correlation coefficient for within each farm-size category was positive and significant for both ends of farm-size classification i.e., marginal and large farm-size categories while non-significant for small, semi-medium and medium farm-size categories. Regression equation model was also applied using operational holdings as independent variable and productivity of the farm as dependent variable. It was observed that the intercept and coefficient were significant and positive. Overall equation was significant and the R-square was very low (4.28%). Farm-size and input-use relationship was positive for fertilizer & manures, plant protection chemicals, hired labour, diesel while negative for hiring machinery. The relationship was found non-significant for repairs & maintenance. The main source of income of farm households was crops (81.89%), followed by regular wages/ salaries (8.67%), livestock (6.61%) and non-farm business (2.32%). The sources of income for marginal farmers are completely different from other farm-size categories. The share of crops income was 39.54 per cent, followed by regular wages/ salaries (26.36%), livestock (20.80%), non-farm business (8.18%) and casual wages (5.12%). Only the marginal farm-size worked as casual workers and earned income as casual wages. There was huge income gap between upper and lower strata income. The distribution of income was highly skewed. It was found that the households in lower strata were mostly marginal and small while the large farm-size categories were mainly concentrated on upper strata. The income sources of the lower quintiles were diversified while that of the upper quintiles was concentrated on agriculture. The Gini coefficient for income inequality among the farm households was 0.50. Using Theil index, it was analyzed that the inequality within farm size categories (0.27) was higher than the inequality between categories (0.18). Further, enquiry into each category revealed that the inequality in all the categories have different sources. The extent of inequality was quite different for within all farm-size categories. The inequality decreased with increase in farm-size except large farm-size holdings. Regular wages came out to be inequality increasing factor in all categories except large category. Livestock came out to be inequality decreasing factor in overall decomposition as well as intra class decomposition of income inequality. This re-emphasis the importance of livestock sector in the agricultural economy of Punjab as well as India (Pavithra and Vatta 2013, Birthal et al 2014, Choudhary and Singh 2019) and at the international level (Adams and He 1995, Lanjouw and Shariff 2004). Dairy sector should be strengthened to enhance the income of farmers particularly marginal and small farmers. It was also suggested that government should facilitate for the adoption of Cooperative farming. Surprisingly, the non-farm business has non-significant impact on the income distribution in Punjab contrary to the finding of Pavithra and Vatta (2013). There is need to explore the possibility to enhance the income from non-farm sources, as this source have been identified as inequality decreasing sources (Adams and He 1995, Lanjouw and Shariff 2004, Pavithra and Vatta 2013, Birthal et al 2014). Thus agro-processing industries settlement is need of hour for provision of non-farm employment opportunities. Focus should be on skill development centres for generation/adoption of self-employment.

Conclusion

To conclude, it was found most of the farm households belonged to general category (84.41%). The farm households with the education level of graduation or above were more prevalent larger farm-size categories. The main crops grown in Punjab were wheat and paddy with around 76 per cent area under these crops. The cropping intensity had positive relationship with farm-size categories. Returns over paid-out cost per hectare had inverted U-shape. There was no inverse relationship between farm-size and productivity in kharif as well as rabi crops. Overall, the absence of inverse farm-size and productivity relationship was re-confirmed for Punjab. Positive relationship between farm-size and productivity was confirmed between marginal farm-size category and all other categories. The researcher was unable to confirm positive relationship for the farm-size categories higher than marginal farms but confirms absence of negative relationship between farm size and productivity in this range. Farm-size and per hectare input-use relationship was positive for fertilizer & manures, plant protection chemicals, hired labour, diesel while negative for hiring machinery. The sources of income for marginal farmers are completely different from rest of the farm-size categories. The marginal and small farm households were mainly engaged as casual workers and earned income as casual wages. The analysis of distribution revealed that there was huge income gap between upper and lower strata income. The households in lower strata were mostly marginal and small farm households. Moreover, the income sources of the lower quintiles were diversified while that of the upper quintiles was concentrated on agriculture showing distress in lower quintiles. The Gini coefficient for inequality among the farm households in Punjab was 0.50 depicting high income inequality among the farm households. Livestock and casual wages were found to be income inequality decreasing factors. The analysis of income inequality within each category revealed that the inequality decreased with increase in farm-size except large farm-size holdings where Gini coefficient was high (0.41). Regular wages came out to be inequality increasing factor in all categories except large category. Livestock came out to be inequality decreasing factor in overall decomposition as well as intra class decomposition of income inequality.

Suggestions and Policy implications

- Government should facilitate for the adoption of cooperative farming for the farm households belonging to marginal and small farm-size categories.
- Livestock sector should be strengthened to enhance the income of farmers particularly marginal and small farmers.
- Work under casual wages can be enhanced under various government like effective implementation of MGNREGA to provide immediate income relief to marginal and small farm households.
- Share of non-farm income among farm households was low. Thus, impetus to agro-processing industries should be provided. It will serve two purpose i) increase income of the farm households engaged in agro-processing ii) non-farm employment opportunities to rural population.
- Focus should be on skill development centres for generation of self-employment opportunities.

References

Adams R H and He J J (1995) Sources of income inequality and poverty in rural Pakistan. Research report 102, International Food Policy Research Institute, Washington DC, USA.

Amirapu A and Subramanian A (2015) Manufacturing or services? An Indian illustration of a development dilemma. Working Paper 409, Issue June 2015, Center for Global Development, Washington DC, USA.

Anonymous (1975) Structural causes of the economic crisis. Econ Pol Wkly 10: 83-86.

Anonymous (2015a) Farmers' income in India: Evidence from secondary data. Agril Sit India 74: 30-70.

Anonymous (2015b) In It Together: Why Less Inequality Benefits All. OECD Publishing, Paris, Retrieved from http://dx.doi.org/10.1787/9789264235120-en on 25.04.2020.

Anonymous (2016a) Income, Expenditure, Productive Assets and Indebtedness of Agricultural Households in India. NSS 70th round, Ministry of Statistics and Programme Implementation, Government of India, New Delhi, India.

Anonymous (2016b) All India Report on Number and Area of Operational Holdings. Agricultural Census Division, Department of Agriculture and Co-operation, Government of India, New Delhi, India.

Anonymous (2018a) Agricultural Statistics at A Glance. Department of Agriculture, Cooperation and Farmers Welfare, Government of India.

Anonymous (2018b) NABARD All India Rural Financial Inclusion Survey 2016-17. National Bank for Agriculture and Rural Development, Mumbai, India.

Anonymous (2019) Agricultural Statistics at a Glance. Department of agriculture, cooperation and farmers welfare, Government of India.

Anonymous (2020) Annual report: Periodic Labour Force Survey 2018-19. Ministry of Statistics and Programme Implementation, National Statistical Office, New Delhi.

Bardhan P K (1973) Size, productivity, and returns to scale: An analysis of farm-level data in Indian agriculture. J Pol Econ 81:1370-86.

Bellu L G and Liberati P (2006) Policy impacts on inequality: Decomposition of income inequality by subgroups. EASYPol, Food and Agriculture Organization, Italy.

Bhalla G S and Chadha G K (1982) Green revolution and the small peasant: A study of income distribution in Punjab agriculture. Econ Pol Wkly 17: 826-33.

Bhalla S S and Roy P (1988) Mis-specification in farm productivity analysis: The role of land quality. Oxford Economic papers 40: 55-73.

Bharadwaj K (1974) Notes on farm size and productivity. Econ Pol Wkly 9: A11-24.

Bhattacharya N and Saini G R (1972) Farm size and productivity: A fresh look. Econ Pol Wkly 7: A63-72.

Birthal P S, Negi D S, Jha A K and Singh D (2014) Income sources of farm households in India: Determinants, distributional consequences and policy implications. Agric Econ Res Rev 27: 37-48.

Chaddha A N (1978) Farm size and productivity revisited: Some notes from recent experience of Punjab. Econ Pol Wkly 32: A82-96.

Chakravorty S, Chandrasekhar S and Naraparaju K (2016) Income generation and inequality in India's agricultural sector: The consequences of land fragmentation. Indira Gandhi Institute of Development Research, Mumbai, India.

Chattopadhyay M and Rudra A (1976) Size-productivity revisited. Econ Pol Wkly 11: A104-16.

Chattopadhyay M and Rudra A (1977) Size-productivity revisited: Addendum. Econ Pol Wkly 12: A476-84.

Choudhary B B and Singh P (2019) How unequal is rural Punjab? Empirical evidence from spatial income distribution. Curr Sci 117: 1855-66.

Cowell F (2005) Theil, ineauality indicies and decomposition. Working paper 1, Society for the Study of Economic Inequality, London, England.

Dasgupta B (1977) India's green revolution. Econ Pol Wkly 12: 241-60.

Deininger K, Jin S, Liu Y and Singh S (2016) Labour market performance and the farm size- productivity relationship in rural India. 26th ICAE Agriculture in Interconnected World. Milan, Italy.

Deolaikar A B (1981) The inverse relationship between productivity and farm size: A test using regression data from India. American J Agri Econ 63: 275-79.

Dhanagare D N (1988) The green revolution and social inequalities in rural India. Bull Conc Asian Scholars 20: 2-13.

Frisvold G B (1994) Does supervision matter? Some hypothesis test using Indian farm level data. J Dev Econ 43: 217-308.

Gahukar R T (1992) Green revolution in food crops: An Indian experience. Outlook on Agric 21: 129-36.

Galor O and Moav O (2004) From physical to human capital accumulation: Inequality and the process of development. The Rev Econ Stud 71: 1001-26.

Ghose A K (1979) Farm size and land productivity in Indian agriculture: A reappraisal. J Dev Stud 16: 27-49.

Gollin D (2018) Farm size and productivity: Lessons from recent literature. FAO, IFAD, ISPC/CGIAR, and the World Bank Expert Consultation: Focusing Agricultural and Rural Development Research and Investment on Achieving SDGs. Retrieved from https://ispc.cgiar.org/sites/default/files/files/events/Joint%20Initiative%202018/Gollin.pdf on 20-03-2020.

Grover D K, Singh J M, Kaur A and Kumar S (2017) State Agricultural Profile-Punjab (AERC Study No. 44). Agro-Economic Research Centre, Punjab Agricultural University, Ludhiana.

Gulati A, Roy R, and Hussain S (2017) Getting Punjab Agriculture Back on High Growth Path: Sources, Drivers and Policy Lessons. Indian Council for Research on International Economic Relations, New Delhi.

Hazell P B R (2009) The Asian Green Revolution. Food Secur Discussion Paper.

Kaur M (1992) Study into economic disparity in the rural sector of Punjab. Ph. D. Dissertation, Punjab Agricultural University, Ludhiana, India.

Kaur M and Sekhon M K (2005) Rice-wheat system and income inequality in rural Punjab. Agri Sit India 759-64.

Kaur S and Singh G (2014) Analysis of income distribution among marginal and small farmers in rural Punjab. International J Sci Res 3:755-59.

Kautsky K (1988) The Agrarian Question. Pp 252-82. Zwan Publications, London and Winchester.

Khusro A M (1964) Returns to scale in Indian agriculture. Indian J Agric Econ 19: 51-80.

Kimhi A, Arayama Y and Kim J M (2006) Identifying determinants of income inequality in the presence of multiple income sources: the case of Korean farm households. EAAE Congress, Agri-food and Rural Innovations for Healthier Societies, Slovenia.

Lamb R L (2002) Inverse productivity: Land quality, labour markets and measurement error. J Dev Econ 71: 71-91.

Lanjouw P and Shariff A (2004) Rural non-farm employment in India: Access, income and poverty impact. Econ Pol Wkly 39: 4429:46.

Lazear E P and Rosen S (1981) Rank-order tournaments as optimum labor contracts. J Pol Econ 89: 841-64.

Lerman R I and Yitzhaki S (1985) Income inequality effects by income sources: A new approach and applications to the United States. The Rev Econ Stat 67: 151-56.

Majid N (2019) Structural change and employment in India. ILO/SIDA Partnership on Employment Working papers, Inernational Labour Organization, Geneva.

Mazumdar D (1965) Size of farm and productivity: A problem of Indian peasant agriculture. Economica 32: 161-73.

Meena M S, Singh K M, Singh R K P, Kumar A, Kumar A and Chahal V P (2017) Inequality and determinants of income among rural households in tribal dominated areas of Jharkhand. Indian J Agric Sci 87: 92-96.

Narayanamoorthy A (2017) Farm income in India: Myths and realities. Indian J Agril Econ 72 1: 48-75.

Overton M (1996) Agricultural Revolution In England: The Transformation of the Agrarian Economy 1500-1850. Cambridge University Press: Cambridge.

Pavithra S and Vatta K (2013) Role of non-farm sector in sustaining rural livelihoods in Punjab. Agric Econ Res Rev 26: 257-65.

Piketty T (2014) Capital in the Twenty-first Century. Belknap Press of Harvard University Press, Cambridge, MA.

Pingali P L (2012) Green revolution: Impacts, limits, andthe path ahead. PNAS 109: 12302–8.

Pisani E (2006) Some socio-economic consequences of the green revolution. Land Reform 2: 97-107.

Ranganathan T, Tripathi A and Rajoriya B (2016) Changing sources of income and income inequality among Indian rural households. National seminar on Dynamics of Rural Labour Relations, National Institute of Rural Development and Panchayati Raj, Hyderabad, India.

Rao A P (1967) Size of holding and productivity. Econ Pol Wkly 2: 1989-91.

Rao C H H (1963) Alternative explanations of the inverse relationship between farm size and output per acre in India. Indian Econ Rev 1: 1-12.

Rudra A (1968a) Farm size and yield per acre. Econ Pol Wkly 3: 1041-44.

Rudra A (1968b) More to returns to scale in Indian agriculture. Econ Pol Wkly 3: 33-38.

Rudra A and Sen A K (1980) Farm size and labour use: Analysis and policy. Econ Pol Wkly 15 :391-94.

Saini G R (1976) Green revolution and the distribution of farm income. Econ Pol Wkly 11: A17-22.

Saini R and Kumar R (2020) Income pattern and consumption behaviour of farm households in Punjab. Agric Res J 57: 923-32

Sen A K (1962) An aspect of Indian agriculture. The Econ Wkly 14: 243-46.

Sen A K (1964) Size of holding and productivity. The Econ Wkly 16: 323-36.

Sengupta and Chattopadhyay (1997) Farm size and productivity- A new look at the old debate. Econ Pol Wkly 32: A172-75

Silber J (ed) (2020) Economic Studies in Inequality, Social Exclusion and Well-Being. Pp 183-86. Springer Nature, Switzerland.

Singh A K (2013) Income and livelihood issues of farmers: A field study in Uttar Pradesh. Proc 21st Agric Econ Res Rev. Vol 26, pp 89-96.

Singh A, Prasanna P A L and Chand R (2011) Farm size and productivity: Understanding the strength of smallholders and improving their livelihoods. Econ Pol Wkly 26/27: 5-11.

Singh I P (1985) Technological changes vis-à-vis income distribution: Inter and intra temporal and spatial analysis in the rural Punjab. Ph.D. Dissertation, Punjab Agricultural University, Ludhiana, India.

Singh K, Grover D K, Vatta K and Kumar S (2000) Changes in structural distribution of land holdings and its impact on productivity of major crops in Punjab. Agro Economic Research Centre, Punjab Agricultural University, Ludhiana.

Singh L and Singh S (2002) Deceleration of economic growth in Punjab: Evidence, explanation and a way-out. Econ Pol Wkly 37: 579-86.

Singh R K P, Kumar A, Singh K M, Chandra N, Bharati R C, Kumar U and Kumar P (2018) Farm size and productivity relationship in smallholder farms: Some empirical evidence from India. J Commun Mobiliza Sustain Dev13(1):61-67.

Singh S, Bhogal S and Singh R (2014) Magnitude and determinants of indebtedness among farmers in Punjab. Indian J Agril Econ 69: 244-56.

Singh S, Sharma V K, Kingra H S and Bhogal S (2013) Income level, expenditure pattern and poverty among Punjab farmers. J Agri Dev Pol 23: 26-39.

Singh S, Kaur M and Kingra H S (2008) Indebtedness among farmers in Punjab. Econ Pol Wkly 43: 130-36.

Stiglitz J (2012) The Price of Inequality: How Today's Divided Society Endangers Our Future. W.W. Norton and company, New York, USA.

Thakur J, Bose M L, Hossain M and Janaiah A (2000) Rural income distribution and poverty in Bihar insights from village studies. Econ Pol Wkly 35: 4653-63.

Todaro MP, Smith SC (2007) Economic Development. Pp 229-291. Pearson Education, USA.

Vatta K (2006) Non-farm employment, poverty and inequality in rural Punjab. Ph.D. Dissertation, Punjab Agricultural University, Ludhiana, India.

Vatta K and Pavithra S (2016) Household income inequality and asset distribution: The case of rural Punjab. In: Singh L and Singh N (eds) Economic transformation of a developing Economy. Pp 215-26. Springer, Singapore.

Vatta K and Sidhu R S (2010) Rural non-farm employment, income distribution and poverty: Micro level evidence from Punjab. Indian J Agril Econ 65: 694-709.

Vatta K, Garg B R and Sidhu M S (2008) Rural employment and Income: The inter-household variations in Punjab. Agric Econ Res Rev 21: 201-10.

Vinutha H P, Poornima B, and Sagar B M (2018) Detection of Outliers Using Interquartile Range Technique from Intrusion Dataset. In: Satapathy S C Tavares J M R S Vikrant Band Mohanty J R (Eds.) Inf Decis Sci, Springer Singapore, 1052-59.

Index

9 789359 191430

Printed at Repro India Ltd.